RAINFALL NOWCASTING MODELS FOR EARLY WARNING SYSTEMS

HYDROLOGICAL SCIENCE AND ENGINEERING

TOMMY S. W. WONG – SERIES EDITOR –
NANYANG TECHNOLOGICAL UNIVERSITY, SINGAPORE

Kinematic-Wave Rainfall-Runoff Formulas
Tommy S. W. Wong
2011. ISBN: 978-1-60692-705-2

Overland Flow and Surface Runoff
Tommy S. W. Wong
2011. ISBN: 978-1-61122-868-7

Rainfall Forecasting
Tommy S. W. Wong
2012. ISBN: 978-1-61942-134-9

Flood Risk and Flood Management
Tommy S. W. Wong
2012. ISBN: 978-1-62081-220-4

HYDROLOGICAL SCIENCE AND ENGINEERING

RAINFALL NOWCASTING MODELS FOR EARLY WARNING SYSTEMS

DAVIDE LUCIANO DE LUCA

Novinka

New York

NOTICE TO THE READER

Library of Congress Cataloging-in-Publication Data
De Luca, Davide Luciano.
  Rainfall nowcasting models for early warning systems / Davide Luciano De Luca
(Department of Soil Conservation, University of Calabria, Italy).
     pages cm
Includes bibliographical references and index.
ISBN 978-1-62257-805-4 (hardcover)
1. Nowcasting (Meteorology) 2. Rain and rainfall--Mathematical models. I. Title.
QC997.75.D42 2013
551.64'77--dc23   2012038026

Published by Nova Science Publishers, Inc. † New York

CONTENTS

Dedicated to
my parents, Pasquale De Luca and Filomena Martorello

PREFACE

This book provides a state of the art overview of rainfall nowcasting or short-term rainfall forecasting, which is an essential component of early warning systems for mitigating risks of landslide and floods. Special attention is given to spatial and temporal hydrological scales that are associated with the forecast of flash floods and shallow-landslides.

The importance of Probabilistic Quantitative Precipitation Forecast (PQPF) is highlighted. This approach is beneficial to a Quantitative Precipitation Forecast (QPF).

Moreover, several approaches for nowcasting are compared; in particular, the differences in performances between stochastic and meteorological models. A methodology for coupling these approaches is proposed, which can improve nowcasting "outside a storm event" (i.e. prior to the commencement of rainfall).

ACKNOWLEDGMENTS

I would like to thank Tommy Wong for his precious suggestions related to the organization of the book.

LIST OF FIGURES

LIST OF TABLES

LIST OF SYMBOLS

a_0, a_1, a_2	coefficients of regression
A, B, C, D, E	events
$\overline{A}$	complement event
$\underline{A}$	array of parameters for regression analysis
$\underline{\underline{A}}_k$	matrix of parameters
$B()$	bivariate standard normal cumulative distribution
$\underline{B}$	array for regression analysis
$\underline{\underline{B}}_k$	matrix of parameters
$a_{jj}^{(k)}$	generic element along the main diagonal of matrix $\underline{\underline{A}}_k$
$b_{jj}^{(k)}$	generic element along the main diagonal of matrix $\underline{\underline{B}}_k$
C	Copula function
$Cov\left[X_i, X_j\right]$	theoretical value of Covariance
$\underline{Cov}$	matrix of covariance
dh	differential of h
c	data
e_{i+1}	error to be minimized
$E[\ ]$	expected value operator
$f_{00|Z_0>1}$	dry ratio

$f_{10\|Z_0>1}$	wet-to-dry ratio
$f_{01\|Z_0>1}$	wet-to- wet ratio
$f_H(h)$	probability density function of H
$f_X(x),\ f_X(x_i)$	probability density function of X
$f_H^{(+)}(h)$	probability density function of H conditional on $H>0$
$f_{H_{i+1}\|Z_i}(h_{i+1}\mid z_i)$	probability density function of H_{i+1}, conditional on a particular value of z_i
$f_{X_1\|X_2}(x_1\mid x_2)$	probability density function of X_1, conditional on a particular value of x_2
$f_{H_{i+1},0}^{(+,0)}(h_{i+1}),\ f_{0,Z_i^{(v)}}^{(0,+)}(z_i^{(v)}),$ $f_{Z_i^{(v)}}^{(+,+)}(z_i^{(v)}),\ f_{H,0,0}^{(+,0,0)}(h)$	univariate density conditional functions
$f_W^{(+,+,0)}(w),\ f_Z^{(+,0,+)}(z)$ $f_{X_1}(x_1)$	univariate marginal density function
$f_{W,Z}^{(+,+,+)}(w,z)$	bivariate marginal density function
$f_{X_1,X_2}(x_1,x_2),$ $f_{H_{i+1},Z_i^{(v)}}(h_{i+1},z_i^{(v)})$	bivariate density function
$f_{H_{i+1},Z_i^{(v)}}^{(+,+)}(h_{i+1},z_i^{(v)}),$ $f_{H\|W,0}^{(+,+,0)}(h\mid w)\ f_{H\|0,Z}^{(+,0,+)}(h\mid z)$	bivariate density conditional function
$f_{H,W,Z}(h,w,z)$	trivariate density function
$f_{H\|W,Z}(h\mid w,z),$ $f_{H\|W,Z}^{(+,+,+)}(h\mid w,z)$	trivariate density conditional function
$F_H(h)$	cumulative density function of H
$F_X(x),F_X(x_i)$	cumulative density function of X
$F_{X_1,X_2}(x_1,x_2)$	bivariate cumulative density function
$F_{X_1\|X_2}(x_1\mid x_2)$	cumulative density function of X_1, conditional on a particular value of x_2

$F_{H|W,Z}(h|w,z),$

$F_{H|W,Z}^{(+,+,+)}(h|w,z)$ — trivariate cumulative density conditional function

$F_{H_{i+1}|Z_i}(h_{i+1}|z_i)$ — cumulative density function of H_{i+1}, conditional on a particular value of z_i

$F_{H_{i+1},Z_i^{(v)}}(h_{i+1},z_i^{(v)}),$ — bivariate cumulative density function

$F_{H|W,0}^{(+,+,0)}(h|w)\ F_{H|0,Z}^{(+,0,+)}(h|z)$

$F_{H_{i+1},0}^{(+,0)}(h_{i+1}),\ F_{0,Z_i^{(v)}}^{(0,+)}(z_i^{(v)}),$ — univariate cumulative density conditional functions

$F_{H_{i+1}}^{(+,+)}(h_{i+1}),\ F_{Z_i^{(v)}}^{(+,+)}(z_i^{(v)}),$

$F_{H_{i+1}|Z_i^{(v)}}(h_{i+1}|Z_i^{(v)}=0),$

$F_{H_{i+1}|Z}(h_{i+1}|Z_i^{(v)}=z_i^{(v)}>0)$

$F_{Z_i^{(v)}|H_{i+1}}(z_i^{(v)}|h_{i+1})$

$F_{H_{i+1}|Z_i^{(v)}}^{(+,+)}(h_{i+1}|z_i^{(v)}),$

$F_{H,0,0}^{(+,0,0)}(h)$

$F_{H_{i+1},Z_i^{(v)}}^{(+,+)}(h_{i+1},z_i^{(v)})$ — bivariate cumulative density conditional function

$F_{H_{i+1}}(h_{i+1}),\ F_{Z_i^{(v)}}(z_i^{(v)})$ — univariate cumulative marginal functions

$f_{\underline{S}_{i+1}|\underline{H}_{i+1},\underline{H}_0}(\underline{s}_{i+1}|\underline{h}_{i+1},\underline{h}_0)$ — likelihood function in the Hydrologic Uncertainty Processor for a Bayesian Forecasting System

$f_{\underline{X}}(\underline{x})$ — multivariate density probability function

$F_{\underline{X}}(\underline{x})$ — multivariate cumulative density function

$_2F_1(a,b;c;u)$ — hypergeometric function

$g(.),g^*(.)$ — generic functions

g_1 — sample value of skewness

g_2 — sample value of kurtosis

$g_{\underline{H}_{i+1}|\underline{H}_0}(\underline{h}_{i+1}|\underline{h}_0)$ — prior density in the Hydrologic Uncertainty Processor for a Bayesian Forecasting System

H	random variable related to the rainfall height (mm)
h	a particular value of H (mm)
H_{i+1}	random variable related to the rainfall height (mm) to be predicted at the future instant i+1
h_{i+1}	a particular value of H_{i+1} (mm)
$\underline{H}_{i+1}$	random vector of rainfall heights (mm)
H_0	statistical null hypothesis rainfall height before the forecasting (mm)
h_0	a particular value of H_0 (mm)
$\underline{H}_0$	vector of rainfall heights before the forecasting (mm)
$\underline{h}_0$	vector of a particular values of $\underline{H}_0$ (mm)
i	temporal interval
I	certain event
k	temporal lag; number of values assumed by a discrete random variable; number of intervals into an histogram
K_C	CDF of a Copula function
L	likelihood function
M	set of model prediction for an ANN
$\underline{\underline{M}}$	matrix of parameters for regression analysis
M_R	sample moment of order R
$m_{H|Z_0>1}$ $m_{H|Z_0=0}$ m_H m_W	mean value
m_Z $m_{y_{op}}$	
m_X	sample value of mean
n	number of Bernoulli experiments; number of sample data
N	number of sites; number of values assumed by a discrete random variable
$NMSE$	Normalized Mean Squared Error
O_n	function evaluated for the nth hidden node of an ANN
$P()$	incomplete gamma function

$P[.]$	probability concerning an assigned event
$P[\underline{w}\,|M]$	prior distribution
$P[D\,|\,\underline{w},M]$	likelihood distribution
$P[\underline{w}\,|\,D,M]$	posterior distribution
p	$P[H>0]$ autoregressive order; number of marginal distribution with unit mean in the Moran-Dowton function; pressure
p_{s0}	pressure value at the surface
p_{top}	pressure value at the top level
$p_0(z_i)$	probability associated to the event $H_{i+1}=0$, as function of z_i
$P_{10}\ \ P_{01}\ \ P_{11}\ \ P_{000}\ \ P_{100}$ $P_{010}\ \ P_{001}\ \ P_{110}\ \ P_{101}\ \ P_{011}$ P_{111}	probabilities
$PP(i)$	i^{th} value of the plotting position
q	moving-average order; standard normal density function
Q	standard normal cumulative distribution
Q^{-1}	inverse function of Q
$r_{1|Z_0>1}\,,\ r_{1|Z_0=0}$	autocorrelation of lag 1
$r_{HW|Z_0>1}$	spatial correlation
$r_{HW}^{(+,+,+)}\,,\ r_{HZ}^{(+,+,+)}\,,\ r_{WZ}^{(+,+,+)}$	sample linear correlation coefficient
r_k	sample autocorrelation coefficient with lag k
R_U	generated random number
$R(\theta_{hwz})$	Function to minimize for evaluation of the association parameter
$s_{H|Z_0>1}\,,\ s_H\,,\ s_W\,,\ s_Z\,,$ $s_{H|Z_0=0}$	standard deviation
S	summation to minimize into the method of the

	ordinary least squares
$\underline{S}_{i+1}$	vector column representing the output of the meteorological model.
$S_1,\ S_2$	activating function of an ANN
$S_p[q]$	summation in the Moran-Dowton distribution
s_X^2	sample value of variance
$t_0,\ t_1,\ t_2, t_N$	temporal instants
T	temperature
u	eastward component of velocity
$\underline{U}$	Vector comprising all the input variables which are affected by uncertainty in a meterological models
v	northward component of velocity
$V_{k,j}$	weights
V_X	variation coefficient of the random variable X
$w_n,\ w_{kn},\ \underline{w}$	connection weights of an ANN
$W_{k,j}$	weights
$W_{i+1}\ W$	random variables (mm) equal to a weighted average of the simultaneous rainfall heights over the four closest neighboring pixels
X, X_1, X_2, X_i, X_j	random variable
x, x_1, x_2, x_i, x_j	particular value of random variable
$x_{\max}$	maximum sample value
$x_{\min}$	minimum sample value
x_{mod}	mode
x_P	percentile, quantile
$x_{0.5}$	median value
$\underline{X}, \underline{X}_1, \underline{X}_2$	vector random variable
$\underline{x}, \underline{x}_1, \underline{x}_2$	particular value of vector random variable
y	transformed rainfall height for an ANN
y_{op}	transformed observed rainfall data for an ANN

y_{op}^{*}	transformed network output for an ANN
Y	random variable
Z_i	random variable as function of antecedent rainfall heights (mm), evaluated at the instant i
z_i	a particular value of Z_i (mm)
$Z_i^{(v)}\ Z_{*i}^{(v)}$	random variable as function of v antecedent rainfall heights (mm), evaluated at the instant i
Z	standardized random variable; random variable as function of v antecedent rainfall heights (mm), evaluated at the instant i
Z_0	Z variable the previous height hours of recorded data
α	average value of Poisson occurrences into the temporal interval; factor of scale for a Weibull distribution; factor of scale for a Pareto distribution; factor of scale for a Log-Logistic distribution; probability associated to the rejection of a true hypothesis (type I error)
$\alpha'_j,\ a,\ b,\ c,\ ,\ \alpha_j,\ \beta'_j$	generic coefficients
β	generic coefficient; average value of the temporal interval between two consecutive Poisson occurrences; factor of shape for a Weibull distribution; factor of shape for a Pareto distribution; factor of shape for a Log-Logistic distribution; probability associated to the non-rejection of a false hypothesis (type II error)
γ_1	theoretical value of skewness
γ_2	theoretical value of kurtosis
$\Gamma(\ )$	complete gamma function
$\delta(.)$	Dirac's function
Δt	temporal resolution
Δx	size of an interval into an histogram
$\Delta x \Delta y$	spatial resolution

ε_{i+1}	white noise (mm)
$\underline{\varepsilon}_{i+1}$	vector of white noise (mm)
$\eta_{\underline{U}}(\underline{u})$	probability density function of a vector $\underline{U}$
$\theta \quad \theta_{hwz} \quad \vartheta$	association parameter
$(\theta_1, \theta_2, ..., \theta_p)$	generic parameters of a probability distribution
$\underline{\theta}$	generic parameter vector of a probability distribution
θ_j	moving-average parameter
$\theta_{l,j}$	moving-average parameters
Θ	residual
λ	parameter of Poisson distribution / parameter of an Exponential distribution
λ_k	spatial lag of autoregressive component
λ_l	spatial lag of moving-average component
$\lambda(z_i)$	function of z_i
$\tilde{\lambda}^{(v+2)}_{1(v+2)}, \ \tilde{\lambda}^{(v+2)}_{11}, \ \tilde{\lambda}^{(v+2)}_{(v+2)(v+2)}$	cofactors of the Laurent edged matrix
$\lambda_h, \eta_h, \lambda_z, \eta_z, \lambda_W, \eta_W$	Parameters of Weibull marginal functions in the Moran Dowton distribution
$\left[\Lambda^{(v+1)}\right]$	Laurent matrix
$\left[\Lambda^{(v+2)}\right]$	Laurent edged matrix
μ	parameter of a normal distribution
μ_H	theoretical mean value of H (mm)
$\mu_X \ \mu_Y$	theoretical mean value of the random variables X, Y
$\underline{\mu}_H$	vector of mean values of $\underline{H}_{i+1}$ (mm)
$\mu_X^{(R)}$	moment of order R for the random variable X
$\overline{\mu_X^{(R)}}$	moment of order R with respect to $E[X]$
$\overline{\overline{\mu_X^{(R)}}}$	moment of order R with respect to the standardized random variable Z
v	number of antecedent rainfall heights

$\phi_{\underline{H}_{i+1}\mid \underline{S}_{i+1},\underline{H}_0}\left(\underline{h}_{i+1}\mid \underline{s}_{i+1},\underline{h}_0\right)$	parameter of Poisson distribution posterior density in the Hydrologic Uncertainty Processor for a Bayesian Forecasting System
ϕ_j	autoregressive parameter
$\phi_{k,j}$	autoregressive parameters
$\varphi(.)$	generator function
$\pi_{\underline{S}_{i+1}}\left(\underline{s}_{i+1}\right)$	output density of the Input Uncertainty Processor for a Bayesian Forecasting System
$\rho_{H_{i+1}H_{i-v+1-m}\cdot H_i\ldots H_{i-v+1}}$	partial autocorrelation
$\rho_{H_{i+1}H_{i-k+1}}$	autocorrelation coefficient
$\rho_{HZ}\quad \rho_{HZ}^{(+,+,+)}\quad \rho_{HW}^{(+,+,+)}$ $\rho_{WZ}^{(+,+,+)}$	theoretical linear correlation coefficient between H and Z
ρ_k	autocorrelation coefficient with lag k
$\rho_{i,j},\ \rho,\ \rho_{X_i,X_j}$	theoretical linear correlation coefficient
$\rho'_{1,0},\rho'_{2,0},\rho'_{3,0},\rho'_{4,0}$	linear correlation coefficients between the reference cell 0 and the neighbor cells
$\underline{\underline{\rho}}$	matrix of linear correlation coefficient
σ	terrain following coordinate parameter of a Normal distribution
$\sigma_X,\ \sigma_y$	theoretical standard deviation for the random variable X and Y
σ_X^2	theoretical variance
$\chi_r(v)$	sample maximum absolute scattering
χ^2	Statistical test
$\psi_{\underline{H}_{i+1}\mid \underline{H}_0}\left(\underline{h}_{i+1}\mid \underline{h}_0\right)$	probability density function in the Integrator Processor for a Bayesian Forecasting System
$\omega_1,\ \omega_2,\ \omega_3$	weights
$\varnothing$	impossible event

Chapter 1

INTRODUCTION

This book focuses on the importance of rainfall nowcasting, or short-term rainfall forecasting (i.e. 1-6 hours forecast in advance), which is an essential component in early warning systems for mitigating risks of landslide and floods.

All related hydrological modelling, such as rainfall-runoff or rainfall-landslide, should be developed with the aim of achieving accurate simulations of real phenomena.

Only with these accurate simulations, it is then possible to forecast landslide and flood events with a sufficiently long lag time so that civil protection measures can be activated.

In the technical literature, many recently developed early warning systems are described. Relating to landslide forecasting, systems operating in Nagasaki (Yano and Senoo 1985), California (Keefer et al. 1987), Rio de Janeiro (d'Orsi et al. 1997), New Zealand (Glade et al. 2000), the UK (Cole and Davis 2002), Hong Kong (Pun et al. 2003; Yu et al. 2004), Washington (Baum, 2005) and Italy (Versace and Capparelli 2008) have been reported. Examples of major operational flood alert systems are:

(1) the real time flood forecasting system named ARTU at the Arno Watershed in Italy (http://www.arno.autoritadibacino.it),
(2) the systems at the Po River Watershed, Italy (Rabuffetti and Barbero 2005),
(3) the Automated Flood Warning System (AFWS) by the US National Weather System (afws.erh.noaa.gov/afws/rss.php), and
(4) the European Flood Alert System (EFAS, Thielen et al. 2009).

Analyzing the structure of an early warning system, it is possible to discriminate four characteristic times (Capparelli and Versace 2011). As shown in Figure 1.1, they are:

1. evolution time, t_1, defined as the time between the phenomenon onset and its impact;
2. lag time, t_2, defined as the time between precursor occurrence and phenomenon onset;
3. nowcasting time, t_3, defined as the time between forecasting and occurrence of the precursor;
4. intervention time, t^*, defined as the time necessary both for making decisions and initiating action such as evacuation and protection of structures and infrastructure.

For phenomena like floods and landslides, rainfall is the main precursor. If $t^* < t_1$ then monitoring the induced phenomenon may be sufficient. If $t_1 < t^* < (t_1 + t_2)$, it is then necessary to measure the rainfall. Finally, if $t^* > (t_1 + t_2)$, rainfall nowcasting is essential. Consequently, for cases where induced phenomena are characterized by a very short response time with respect to precipitation input, such as flash floods and shallow landslides, real-time measurement of precipitation is not sufficient. In these cases, rainfall nowcasting is important so as to give a long enough lag time to activate civil protection measures.

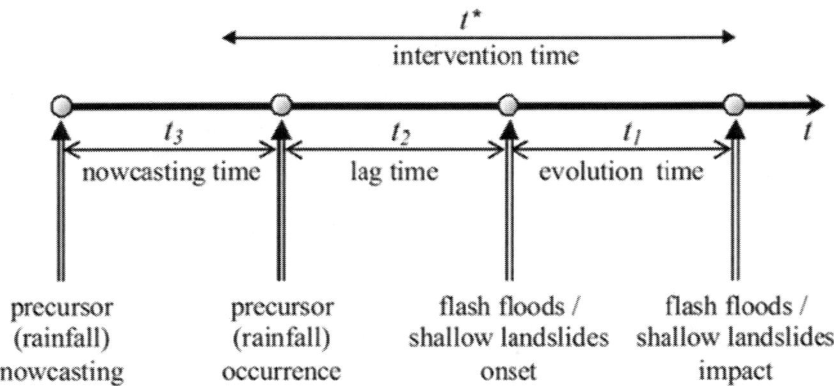

Figure 1.1. Characteristic times into an early warning system (adapted from Capparelli and Versace 2011).

In this context, Chapter 2 describes two approaches in short-term forecasting: (1) deterministic and (2) probabilistic. A deterministic forecast specifies a single estimate of the predicted rainfall as part of a future single time series or a space-time series. A probabilistic forecast specifies a probability distribution of the rainfall. Since the mechanism of rainfall generation is both dynamic and complex, a forecaster should prefer a probabilistic prediction. Only with such a probabilistic approach, it is then possible for a forecaster to quantify the risk associated with induced phenomena, thereby making a rational decision.

Regarding the types of rainfall nowcasting tools, the technical literature proposed stochastic and meteorological models. Stochastic models provide reliable predictions over small temporal and spatial scales, which are of interest in hydrological applications. Depending on the spatial scale, temporal or space-temporal models may be used. Temporal models are adopted for small basins, where uniformly distributed precipitation can be assumed. In medium size basins, as spatial variability of precipitation becomes important, space-time models are required. However, stochastic models are Markovian processes, i.e. they depend on antecedent (rainfall) data only, so they can only exhibit reliable results in the case of nowcasting "inside a storm event" (i.e. after the commencement of a rainfall event).

On the other hand, meteorological models only provide valid qualitative-quantitative rainfall forecasting tools for 24-72 hours. At these forecasting horizons, an absolute precise forecast is not required, but rather an order of magnitude forecast is sufficient. Since these models are based on atmospheric phenomena developing on a synoptic scale, in general especially for complex orography, they are unable to provide a reliable forecast for small temporal and spatial scales, which are of course necessary in hydrological applications. For these reasons, in the case of forecasting "outside a storm event" (i.e. prior to the commencement of rainfall), coupling stochastic models with output results from meteorological models is the best option to improve rainfall nowcasting on hydrological scales.

With the aim of providing the scientific background to readers, Chapter 3 describes several stochastic models that are suitable for rainfall nowcasting and reported in the technical literature. They can be grouped into two main classes:

(1) the model-driven class (see Section 3.1), in which the identification of the type of relationships among the variables (model identification) and then the estimation of model parameters are required.

(2) the data-driven class (see Section 3.2), comprising all the models which depend on the available data to be "learned", without a priori hypothesis about the kind of relationships.

Chapter 4 provides an overview of meteorological models that are suitable for a numerical weather prediction (NWP). Chapter 5 describes the methodology, which can be adopted to couple stochastic and meteorological models, and provides an example of the application. Chapter 6 contains conclusions and guidelines that are useful for both users and developers of early warning systems.

Finally, Appendixes A and B give an overview of the theory of probability distributions and of statistical inference.

Chapter 2

Deterministic and Probabilistic Forecasting

The necessity of forecasting is caused by the uncertainty of future events. In general, a forecast does not eliminate the uncertainty but only reduces it. As mentioned in Chapter 1, there are two approaches in forecasting: (1) deterministic, and (2) probabilistic. Moreover, in the case of rainfall forecasting, the deterministic approach is called Quantitative Precipitation Forecast (QPF), while the probabilistic approach is called Probabilistic Quantitative Precipitation Forecast (PQPF).

A deterministic forecast induces an illusion of certainty. Hence, if the forecast is inaccurate, the consequences can be grave both in terms of economic and social losses. Consequently, a PQPF is preferred as it offers many advantages (Krzysztofowicz 2001). First, it allows the total uncertainty related to the occurrence of a future event to be quantified. As such, the degree of risk associated with civil protection measures can be assigned thereby enabling rational decisions to be made by the authority.

This chapter provides a brief overview of a recommended methodology for probabilistic forecasting. A numerical example is also included so as to illustrate the use of the methodology.

2.1. Mathematical Background of PQPF

Numerical quantification of the total uncertainty related to an occurrence of a future event (in this case a rainfall event) is called "predictive probability", which must obey the laws of probability theory. Therefore, a predictive probability distribution function is required in order to quantify the total uncertainty of the rainfall values (Murphy et al. 1985; Krzysztofowicz et al. 1993; Antolik 2000; Damrath et al. 2000; Golding 2000; Seo et al. 2000). The predictive probability distribution function is conditional on all information available in the forecasting process. The sources of uncertainty may be numerous as they depend on the event being forecasted and the method of forecasting.

In general, Montanari et al. (2009) indicated that they include: (1) inherent randomness, (2) errors in models, (3) errors in model parameters values, and (4) errors in measured data.

In recent years, research activities have been focused on uncertainty quantification. Numerous contributions are concerned with the development of inference methodologies that treat the sources of error separately.

Many of these relied on the classic Bayesian statistics as a framework to integrate the sources of uncertainty (see Chapter 5; Krzysztofowicz 1999, 2002; Kavetski et al. 2006a,b; Kuczera et al. 2006; Liu and Gupta 2007; Huard and Mailhot 2008; Renard et al. 2010).

On the other hand, there are some who used the informal Bayesian approach, such as the popular Generalized Likelihood Uncertainty Estimation or GLUE approach (Beven and Binley 1992).

The predictive probability distribution function is expressed in terms of a probability density function (PDF) and a cumulative density function (CDF) (see Appendix A for further details about the meaning of these concepts). To facilitate the understanding of these concepts, Figure 2.1 shows an example of the probabilistic prediction for the rainfall height H, which is related to a forecast temporal interval Δt at an assigned site (or area) of a spatial domain.

It must be highlighted that H denotes a random variable, while h is the particular value (realization) of H.

In this book, this mathematical symbolism is adopted for all random variables. In this example, H is considered as a mixed random variable. In an analysis of rainfall data on a hydrological temporal scale, (i.e. from a few minutes to a few hours), the data show a high percentage of zero values. Thus, the statistical behavior of such data is correctly modelled by mixed distributions, characterized by a finite probability for rainfall heights equal to zero and an infinitesimal probability for positive values (Kedem et al 1990; Kayano and Shimizu 1994; Kottegoda and Rosso 1997; Yoo et al. 2005).

With this assumption, the probability density function (PDF) $f_H(h)$ is defined as:

$$f_H(h) = (1-p) \cdot \delta(h) + p \cdot f_H^{(+)}(h) \tag{2.1}$$

where $f_H(h) \cdot dh = P[h \leq H \leq h + dh]$, $P[.]$ denotes the probability concerning an assigned event indicated in the squared brackets, $p = P[H > 0]$, $\delta(.)$ is the Dirac's function and $f_H^{(+)}(h)$ is the PDF conditional on $H > 0$, i.e. $f_H^{(+)}(h) \cdot dh = P[h \leq H \leq h + dh \,|\, H > 0]$. The CDF $F_H(h)$ takes the form:

$$F_H(h) = P[H \leq h] = (1-p) + p \cdot F_H^{(+)}(h) \tag{2.2}$$

where $F_H^{(+)}(h) = P[H \leq h \,|\, H > 0]$. The relationships between $f_H(h)$ and $F_H(h)$ are:

$$F_H(h_1) = \int_0^{h_1} f_H(h) \cdot dh \tag{2.3}$$

$$f_H(h) = \frac{dF_H(h)}{dh} \tag{2.4}$$

It must be pointed out that $f_H(h)$ and $F_H(h)$ are predictive probability functions conditional on all available data at the time of forecast. Consequently, predictive uncertainty evolves in time and $f_H(h)$ and $F_H(h)$ must be updated based on the available information.

If H_{i+1} is defined as the rainfall height accumulated over the interval $[i\Delta t; (i+1)\Delta t]$, which is to be forecasted, and Z_i is defined as a function of some previous observed rainfall heights prior to the interval $[(i-1)\Delta t; i\Delta t]$, then the expression of $f_H(h)$ and $F_H(h)$ can be rewritten as $f_{H_{i+1}|Z_i}(h_{i+1}\,|\,z_i)$ and $F_{H_{i+1}|Z_i}(h_{i+1}\,|\,z_i)$, as follows:

$$f_{H_{i+1}|Z_i}(h_{i+1}\,|\,z_i)dh = P[h_{i+1} \leq H_{i+1} \leq h_{i+1} + dh_{i+1}\,|\,Z_i = z_i] \tag{2.5}$$

$$F_{H_{i+1}|Z_i}(h_{i+1}\,|\,z_i) = P[H_{i+1} \leq h_{i+1}\,|\,Z_i = z_i] \tag{2.6}$$

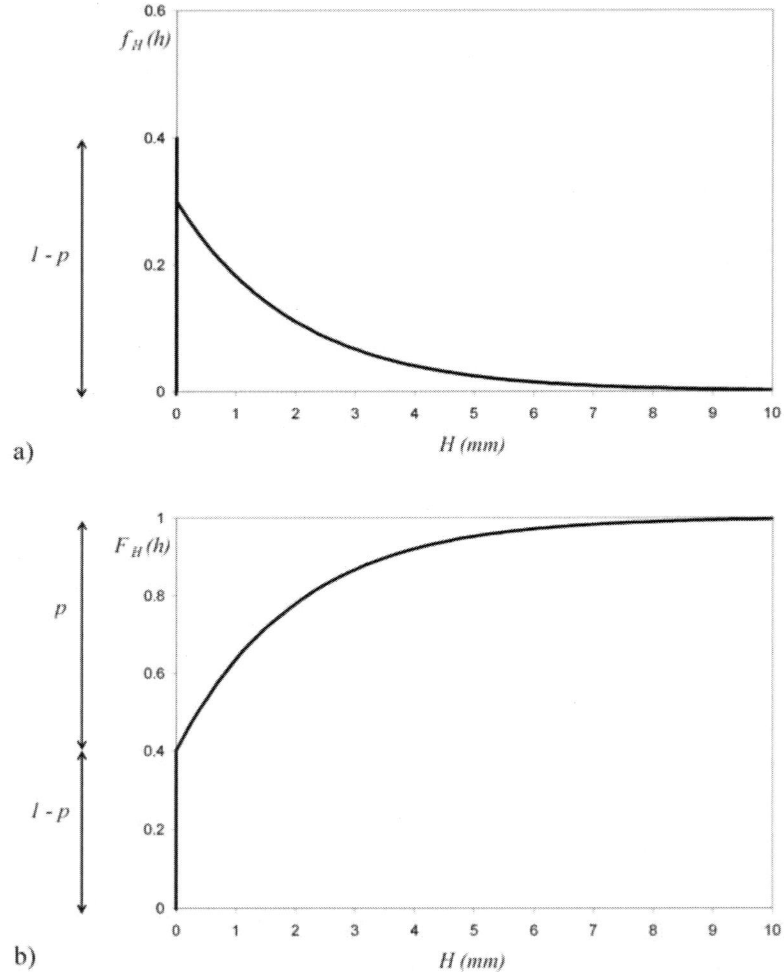

Figure 2.1. Qualitative example of $f_H(h)$ and $F_H(h)$ mixed distributions.

2.2. NUMERICAL EXAMPLE OF PQPF

In the following numerical example, $F_{H_{i+1}|Z_i}(h_{i+1}\mid z_i)$ is defined as:

$$F_{H_{i+1}|Z_i}(h_{i+1}\mid z_i) = p_0(z_i) + \left[1 - p_0(z_i)\right]\left[1 - e^{-\lambda(z_i)h_{i+1}}\right] \qquad (2.7)$$

where:

$$z_i = \frac{1}{3}\sum_{j=1}^{3} h_{i-j+1} \qquad (2.8)$$

$$p_0(z_i) = e^{-z_i} \qquad (2.9)$$

$$\lambda(z_i) = 1/z_i \qquad (2.10)$$

Considering the rainfall data set reported in Table 2.1, the evaluation of CDF for several values of i as starting points of nowcasting can be carried out. As the estimation of z_i requires three values of rainfall height (Eq. (2.8)), all the calculations must be developed from $i=3$, which constitutes the first starting point of nowcasting. Consequently, when $i=3$ the probability distribution is related to the rainfall height of $i=4$, and so on.

Figure 2.2 shows the probability functions for several values of i as the starting point of nowcasting (see Section 3.2 for the nowcasting procedure concerning time intervals greater than $i+1$). Moreover, Figure 2.3 shows an output that can be provided by a developer of rainfall nowcasting models: a comparison of the observed rainfall heights (Table 2.1) and their corresponding percentiles (i.e. the values which should not be exceeded with an assigned probability; see Appendix A for further details).

Table 2.1. Example of calculation for $F_{H_{i+1}|Z_i}(h_{i+1}|z_i)$

i	h_i (mm)	z_i (mm)	$p_0(z_i)$	$\lambda(z_i)$
1	0.0			
2	0.0			
3	3.5	1.17	0.31	0.86
4	3.0	2.17	0.11	0.46
5	5.5	4.00	0.02	0.25
6	5.7	4.73	0.01	0.21
7	6.0	5.73	0.00	0.17
8	7.0	6.23	0.00	0.16
9	5.0	6.00	0.00	0.17
10	0.0	4.00	0.02	0.25
11	2.0	2.33	0.10	0.43

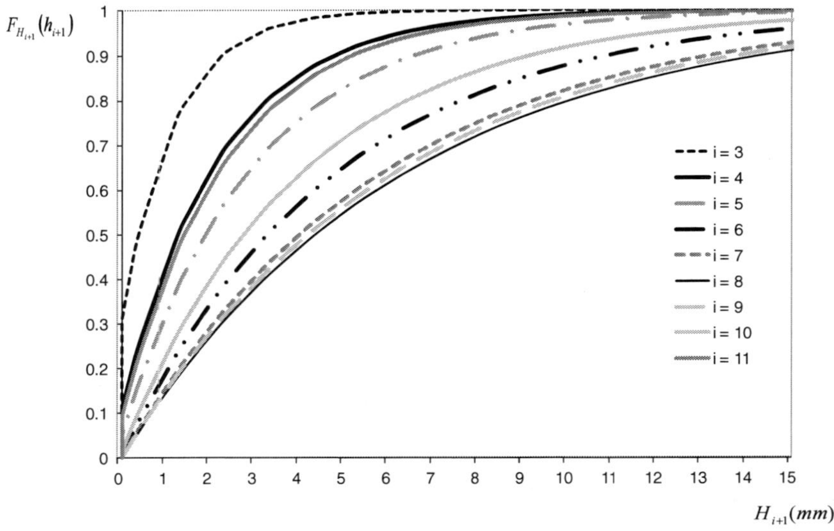

Figure 2.2. Evaluation of $F_{H_{i+1}|Z_i}\left(h_{i+1}\mid z_i\right)$ for different values of starting point i for nowcasting (see Table 2.1).

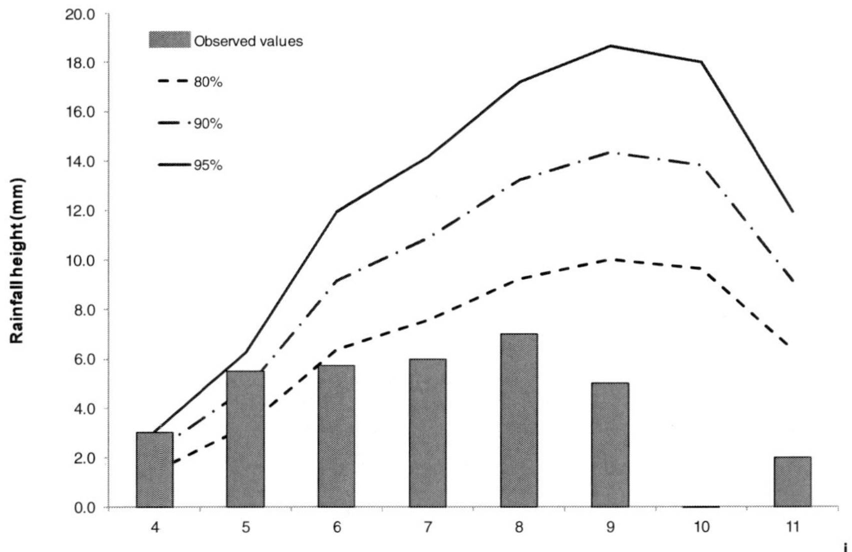

Figure 2.3. Comparison of observed rainfall heights with their corresponding 80%, 90% and 95% percentile forecasts.

In this example, given a probability of non-exceedance, the percentile corresponding to $F_{H_{i+1}|Z_i}\left(h_{i+1} \mid z_i\right)$ can be evaluated by the inversion of Eq. (2.7):

$$h_{i+1} = -\frac{1}{\lambda(z_i)}\ln\left[1 - \frac{F_{H_{i+1}|Z_i}\left(h_{i+1} \mid z_i\right) - p_0(z_i)}{1 - p_0(z_i)}\right] \qquad (2.11)$$

In general, a rainfall nowcasting model is considered well-structured if the future observed values are less than the forecasted values of high non-exceedance probabilities (as shown in this example).

CONCLUSION

This chapter focused on the importance of probabilistic nowcasting for rainfall occurrence. An overview of the methodology is described. This information constitutes the first step of a more general theory, which concerns the use of stochastic models (see Chapter 3) and the coupling with output of meteorological models (see Chapter 5).

Chapter 3

STOCHASTIC MODELS

This chapter contains the scientific background of several stochastic models that have been proposed in technical literature and are suitable for rainfall nowcasting. They provide PQPF predictions. Consequently, if their output is used as input to rainfall-runoff models or models for landslides triggered by rainfall, it is possible to evaluate in real time the probability of occurrence of all hydrological induced phenomena.

Scientific literature contains many approaches aimed to model the characteristics of rainfall processes. First of all, methodologies related to the cluster structure (Austin and Houze 1972) and the theory of point processes (Cox and Isham 1980; Rodriguez-Iturbe et al. 1984) must be mentioned. These models are able to reproduce storm events that are randomly located in space and time.

They adopt Poisson processes for modelling the occurrences of rainfall events, and they can reproduce the shape of a rainfall event ranging from a simple rectangular pulse scheme to cluster models. The former combination of Poisson occurrences/rectangular shape provides the Poisson Rectangular Pulse (PRP) model. The latter combination comprises the Neyman-Scott White Noise (NSWN), the Neyman-Scott Rectangular Pulse (NSRP), or the Bartlett-Lewis Rectangular Pulses (BLRP).

These models have been widely used to reproduce the rainfall features at various scales of aggregation, such as cluster dependence and extreme value properties (Rodriguez-Iturbe et al. 1984, 1987a,b; Rodriguez-Iturbe 1986; Entekhabi et al. 1989; Cadavid et al. 1992; Burlando and Rosso 1993). Unfortunately, their formulation is too complex for nowcasting (Ramirez and Bras 1985). Hence, there is a need for alternative models. One such alternative are the black-box models, which are based on the storm tracking concept (Phanartzis 1979; Nguyen et al. 1978; Johnson and Bras 1980). However, these models are unable to reproduce completely the variability of observed rainfall at several temporal scales (Zawadzki 1987; Burlando et al. 1993). Therefore, other types of models must be used for nowcasting.

The following sections provide a brief overview of two suitable classes of models: (1) the model-driven class, and (2) the data-driven class.

A model of the former class identifies the type of relationships among the variables (model identification) and then estimates the values of the model parameters. This class includes the classical AutoRegressive Moving Average or ARMA models (Section 3.1), and models that are based on mixed distributions, which are able to reproduce the rainfall data at the required hydrological temporal scales (Section 3.2).

The data-driven class comprises all the models that depend on the available data to be "learned", without any "a priori" hypothesis about the relationships among the variables. Artificial Neural Networks (ANNs) belong to this class (Section 3.3).

Regarding the evaluation of model parameters, there are two methodologies (Toth et al. 2000): (1) split-sample calibration, and (2) adaptive calibration. In the split-sample calibration, the historical database is divided into two sets: a calibration set, and a validation set.

The first set should contain at least twice the number of events as compared to the second set. This is to ensure that the process of parameter estimation is robust. The validation set is then used to check the model performance. Further, the two sets of data must have the same proportion of short duration–high intensity and long duration–low intensity rainfall events. Hence, the calibration set would comprise a variety of storm events, which ensures a robust calibration. Moreover, due to their probabilistic nature, the models can forecast, in a stochastic way, rainfall intensities higher than those in the calibration set.

In the adaptive calibration, data of past events are not used. Instead, only the rainfall data immediately preceding the forecast instant (e.g. data of some antecedent hours) are used.

In this way, although the data set may be poor, an on-line recalibration of a model can be carried out. At each time step, as soon as new observations are available, it is then possible to capture the characteristics of the current meteorological event thereby making a forecast into the future. Also in this case, the models are able to forecast, in a stochastic way, rainfall intensities higher than those measured in real time.

3.1. ARMA MODELS

AutoRegressive Moving Average models (ARMA) are commonly used in hydrology to forecast rainfall. In the case of temporal analysis, they are denoted as *ARMA(p,q)*, where p and q represent, respectively, the autoregressive and moving-average orders (Box and Jenkins 1976; Brockwell and Davis 1987; Bras and Rodriguez-Iturbe 1994). Each observation of the time series is modelled as a weighted sum of p previous data, and *(q+1)* values of a white noise process:

$$H_{i+1} = \sum_{j=1}^{p} \phi_j \cdot \left(H_{i+1-j} - \mu_H\right) + \varepsilon_{i+1} + \sum_{j=1}^{q} \theta_j \cdot \varepsilon_{i+1-j} + \mu_H \qquad (3.1)$$

where H_{i+1} is the rainfall height accumulated over the interval $\left[i\Delta t;(i+1)\Delta t\right]$, ε_{i+1} is a white noise (i.e. a zero mean random variable that is not correlated with the past values of H_{i+1}) related to the interval $\left[i\Delta t;(i+1)\Delta t\right]$, ϕ_j and θ_j are respectively the autoregressive and moving-average parameters, and μ_H is the mean of the time series.

The model defined in Eq. (3.1) can be used only for stationary processes with normal marginal distributions. Rainfall processes, however, are typically non-stationary and skewed. Hence, the model cannot be applied directly to observed rainfall data. To overcome the problem of non-stationarity, rainfall data can be grouped by months or seasons. In this way, the model can then be applied to data of a given month or a given season. Moreover, to resolve the problem of skewness, a data transformation, like the Box-Cox approach (Box and Cox 1964), can be used to convert the observed data into a normally distributed dataset.

In general, for the identification of p and q orders, the procedures suggested by Box and Jenkins (1976) and Hipel et al. (1977) can be adopted. In the case of rainfall nowcasting, Obeysekera et al. (1987) demonstrated that the correlation structure derived from PRP, NSWN and NSRP models can be considered equivalent to that of the low-order ARMA process. For example, PRP and NSWN processes and an ARMA(1,1) model present equivalent correlation structures. Similarly, NSRP process and an ARMA(2,2) model present equivalent correlation structures. Moreover, Entekhabi et al. (1989) and Burlando and Rosso (1993) suggested that the NSRP model is the best model to reproduce the stochastic structure of short-term rainfall. Hence, an ARMA(2,2) can be adopted for nowcasting.

3.1.1. Parameter Estimation of ARMA (2,2)

For a given season (in order to respect the hypothesis of stationary process), calibration of ARMA (2,2) models for rainfall nowcasting at high-resolution temporal scale (hourly or sub-hourly) is more difficult than the case of temporal resolutions like months or seasons. This is due to the intermittency of rainfall heights when the data are in such a fine temporal resolution.

Considering the split-sample calibration, two different approaches can be used (Burlando et al. 1993). The first is to take the entire data set, including zero measured rainfall heights. The second is to take only nonzero precipitation (storm) events that are selected from each season, and then arranged sequentially for estimation.

In this second procedure, an objective criterion is required to define a storm from observed data. For example, Burlando et al. (1993) assumed that two nonzero rainfall records belong to two different storms if they are separated by an arbitrary number of hours with zero rainfall.

They used separation periods of 12 hours for winter storms, and 4 hours for summer storms. The difference in the separation period is due to the different rainfall processes during the winter and summer months. Parameter estimation is based on the least square method (Box and Jenkins 1976), in which the parameters ϕ_1, ϕ_2, θ_1 and θ_2 are evaluated by minimizing the sum of the residual square $\sum \varepsilon_{i+1}^2$, where ε_{i+1} is the white noise of Eq. (3.1). However, the results derived from this method may be not satisfactory (Burlando et al. 1993) for both the complete data set (i.e. including dry periods), and the data set with nonzero rainfall only. This is due to the different processes of rainfall generation within the seasons.

A method, which can produce better model performance, is the adaptive calibration. However, there are still problems of biased rainfall evaluation (i.e. only nonzero rainfall are considered), and the intermittence characteristic of precipitation which cannot be modelled. Hence, models that adopt mixed distributions for rainfall (described in Section 3.2.1) are preferred, as they are more "physically" based.

3.1.2. Multivariate ARMA Models

Rainfall nowcasting at several sites requires a procedure that is more involved than that for a single site. Vectors and matrices are needed for analyses of multiple series. A multiple time series $\underline{H}_{i+1}$ is considered, which is constituted by a vector column containing elements $H_{i+1,1}, H_{i+1,2}, ..., H_{i+1,N}$. N is the number of series, corresponding to the number of sites or rain gauges. A multivariate Space-Time ARMA, denoted as $STARMA(p,q)$ (Cliff et al. 1975; Martin and Oeppen 1975; Aronian 1980; Pfeifer and Deutsch 1980; Abraham 1983; Cressie 1993), is defined as:

$$\underline{H}_{i+1} = \sum_{k=1}^{p}\sum_{j=1}^{\lambda_k} \phi_{k,j} \cdot W_{k,j} \cdot \left(\underline{H}_{i+1-k} - \underline{\mu}_H\right) + \sum_{l=1}^{q}\sum_{j=1}^{\lambda_q} \theta_{l,j} \cdot V_{k,j} \cdot \underline{\varepsilon}_{i+1-l} + \underline{\varepsilon}_{i+1} + \underline{\mu}_H$$

$$(3.2)$$

where $\underline{\mu}_H$ is the vector of the mean values related to the N sites, $W_{k,j}$ and $V_{k,j}$ are weights, λ_k and λ_l are respectively the spatial lag of autoregressive and moving-average component. $\underline{\varepsilon}_{i+1}$ is the vector of white noises for the N sites. Each element of $\underline{\varepsilon}_{i+1}$ has zero mean, and it is independent and identically normally distributed. $\phi_{k,j}$ and $\theta_{l,j}$ are respectively autoregressive and moving-average parameters.

Denoting $\sum_{j=1}^{\lambda_k} \phi_{k,j} \cdot W_{k,j}$ and $\sum_{j=1}^{\lambda_l} \theta_{l,j} \cdot V_{l,j}$ as $\underline{\underline{A}}_k$ and $\underline{\underline{B}}_k$ respectively, Eq. (3.2) becomes:

$$\underline{H}_{i+1} = \sum_{k=1}^{p} \underline{\underline{A}}_k \cdot \left(\underline{H}_{i+1-k} - \underline{\mu}_H\right) + \sum_{l=1}^{q} \underline{\underline{B}}_k \cdot \underline{\varepsilon}_{i+1-l} + \underline{\varepsilon}_{i+1} + \underline{\mu}_H \qquad (3.3)$$

where $\underline{\underline{A}}_k$ and $\underline{\underline{B}}_k$ are N by N matrices of parameters.

Using a STARMA model often leads to complex parameter estimation. Hence, simplifications are required. For instance, if $\underline{\underline{A}}_k$ and $\underline{\underline{B}}_k$ are assumed to be diagonal matrices, the resultant model is called the Contemporaneous

AutoRegressive Moving Average or CARMA model (Camacho et al. 1985; Salas et al. 1980, 1985; Burlando et al. 1996), in which there is only the dependence on the component of $\underline{H}_{i+1}$ located at the same point. The model can then be decoupled into their components, as follows:

$$H_{i+1,j} = \sum_{k=1}^{p} a_{jj}^{(k)} \cdot \left(H_{i+1-k,j} - \mu_{H,j} \right) + \sum_{l=1}^{q} b_{ii}^{(k)} \cdot \varepsilon_{i+1-l,j} + \varepsilon_{i+1,j} + \mu_{H,j} \quad (3.4)$$

with $j = 1,...,N$. Thus, the model components at each site are simply univariate $ARMA(p,q)$ models. Consequently, the parameters $a_{jj}^{(k)}$ and $b_{jj}^{(k)}$ in each model can be estimated by using univariate estimation procedures. Moreover, each $\varepsilon_{i+1,j}$ is uncorrelated, but they are contemporaneously correlated with a variance-covariance matrix.

For STARMA and CARMA models, considerations similar to those of correspondent temporal models are applicable. Even though the adaptive calibration may provide a better performance model, the need of an unbiased rainfall forecast induces a preference to use a model that is based on mixed distributions (Section 3.2.2)

3.2. MODEL-DRIVEN APPROACHES BASED ON MIXED DISTRIBUTIONS

This section describes the methodologies by considering the mixed nature of rainfall data (Eqs. 2.1-2.2). The methodologies can be adopted for rainfall nowcasting at a single site (temporal models, Section 3.2.1) or at several sites (spatiotemporal models, Section 3.2.2). Joint probability densities are used, with a structure characterized by a combination of univariate and multivariate probability functions.

Several approaches can be used to model the multivariate distributions. Three approaches are considered in this chapter:

(1) the Moran-Downton multivariate exponential distribution (Section 3.2.3);
(2) the Meta-Gaussian model (Section 3.2.4);
(3) Copula functions (Section 3.2.5).

3.2.1. Temporal Models

In the case of temporal rainfall nowcasting, it is possible to develop another type of Markovian process by considering the following random variables:

(1) Let H_{i+1} be the rainfall height to be forecasted over the interval $[i\Delta t;(i+1)\Delta t]$.

(2) Let $Z_i^{(v)}$ be a linear or nonlinear function of v number of antecedent rainfall heights $H_i, H_{i-1}, ..., H_{i-v+1}$. It is obvious that $Z_i^{(v)}$ is a random variable characterized by a mixed nature similar to H_{i+1}.

With these two assumptions, a bivariate density function $f_{H_{i+1}, Z_i^{(v)}}\left(h_{i+1}, z_i^{(v)}\right)$, such that:

$$f_{H_{i+1}, Z_i^{(v)}}\left(h_{i+1}, z_i^{(v)}\right) \cdot dh_{i+1} \cdot dz_i^{(v)} =$$
$$P\left[h_{i+1} \le H_{i+1} \le h_{i+1} + dh_{i+1} \cap z_i^{(v)} \le Z_i^{(v)} \le z_i^{(v)} + dz_i^{(v)}\right] \quad (3.5)$$

can then be written as:

$$f_{H_{i+1}, Z_i^{(v)}}\left(h_{i+1}, z_i^{(v)}\right) =$$
$$\left(1 - p_{10} - p_{01} - p_{11}\right) \cdot \delta\left(h_{i+1}\right) \cdot \delta\left(z_i^{(v)}\right) + p_{10} \cdot f_{H_{i+1},0}^{(+,0)}\left(h_{i+1}\right) \cdot \delta\left(z_i^{(v)}\right) +$$
$$p_{01} \cdot f_{0,Z_i^{(v)}}^{(0,+)}\left(z_i^{(v)}\right) \cdot \delta\left(h_{i+1}\right) + p_{11} \cdot f_{H_{i+1},Z_i^{(v)}}^{(+,+)}\left(h_{i+1}, z_i^{(v)}\right) \quad (3.6)$$

where:

$$p_{10} = P\left[H_{i+1} > 0 \cap Z_i^{(v)} = 0\right] \quad (3.7)$$

$$p_{01} = P\left[H_{i+1} = 0 \cap Z_i^{(v)} > 0\right] \quad (3.8)$$

$$p_{11} = P\left[H_{i+1} > 0 \cap Z_i^{(v)} > 0\right] \quad (3.9)$$

$$f_{H_{i+1},0}^{(+,0)}\left(h_{i+1}\right)\cdot dh_{i+1} = P\left[h_{i+1} \leq H_{i+1} \leq h_{i+1} + dh_{i+1} \mid H_{i+1} > 0 \cap Z_i^{(v)} = 0\right]$$

$$(3.10)$$

$$f_{0,Z_i^{(v)}}^{(0,+)}\left(z_i^{(v)}\right)\cdot dz_i^{(v)} = P\left[z_i^{(v)} \leq Z_i^{(v)} \leq z_i^{(v)} + dz_i^{(v)} \mid H_{i+1} = 0 \cap Z_i^{(v)} > 0\right]$$

$$(3.11)$$

$$f_{H_{i+1},Z_i^{(v)}}^{(+,+)}\left(h_{i+1},z_i^{(v)}\right)\cdot dh_{i+1} \cdot dz_i^{(v)} =$$

$$P\left[h_{i+1} \leq H_{i+1} \leq h_{i+1} + dh_{i+1} \cap z_i^{(v)} \leq Z_i^{(v)} \leq z_i^{(v)} + dz_i^{(v)} \mid H_{i+1} > 0 \cap Z_i^{(v)} > 0\right]$$

$$(3.12)$$

and its corresponding bivariate cumulative function can be written as:

$$F_{H_{i+1},Z_i^{(v)}}\left(h_{i-1},z_i^{(v)}\right) = P\left[H_{i+1} \leq h_{i+1} \cap Z_i^{(v)} \leq z_i^{(v)}\right] =$$

$$\left(1 - p_{10} - p_{01} - p_{11}\right) + p_{10} \cdot F_{H_{i+1},0}^{(+,0)}\left(h_{i+1}\right) +$$

$$p_{01} \cdot F_{0,Z_i^{(v)}}^{(0,+)}\left(z_i^{(v)}\right) + p_{11} \cdot F_{H_{i+1},Z_i^{(v)}}^{(+,+)}\left(h_{i+1},z_i^{(v)}\right)$$

$$(3.13)$$

with

$$F_{H_{i+1},0}^{(+,0)}\left(h_{i+1}\right) = P\left[H_{i+1} \leq h_{i+1} \mid H_{i+1} > 0 \cap Z_i^{(v)} = 0\right]$$

$$(3.14)$$

$$F_{0,Z_i^{(v)}}^{(0,+)}\left(z_i^{(v)}\right) = P\left[Z_i^{(v)} \leq z_i^{(v)} \mid H_{i+1} = 0 \cap Z_i^{(v)} > 0\right]$$

$$(3.15)$$

$$F_{H_{i+1},Z_i^{(v)}}^{(+,+)}\left(h_{i+1},z_i^{(v)}\right) = P\left[H_{i+1} \leq h_{i+1} \cap Z_i^{(v)} \leq z_i^{(v)} \mid H_{i+1} > 0 \cap Z_i^{(v)} > 0\right]$$

$$(3.16)$$

The bivariate distribution in Eq. (3.13) gives rise to the following distribution functions:

(1) the marginal distributions of H_{i+1} and $Z_i^{(v)}$ are:

$$F_{H_{i+1}}(h_{i+1}) = P[H_{i+1} \leq h_{i+1}] = (1 - p_{10} - p_{11}) + p_{10} \cdot F_{H_{i+1},0}^{(+,0)}(h_{i+1}) + p_{11} \cdot F_{H_{i+1}}^{(+,+)}(h_{i+1})$$

$$(3.17)$$

$$F_{Z_i^{(v)}}\left(z_i^{(v)}\right) = P\left[Z_i^{(v)} \leq z_i^{(v)}\right] = (1 - p_{01} - p_{11}) + p_{01} \cdot F_{0,Z_i^{(v)}}^{(0,+)}\left(z_i^{(v)}\right) + p_{11} \cdot F_{Z_i^{(v)}}^{(+,+)}\left(z_i^{(v)}\right)$$

$$(3.18)$$

where

$$F_{H_{i+1}}^{(+,+)}(h_{i+1}) = P\left[H_{i+1} \leq h_{i+1} \mid H_{i+1} > 0 \cap Z_i^{(v)} > 0\right] \qquad (3.19)$$

$$F_{Z_i^{(v)}}^{(+,+)}\left(z_i^{(v)}\right) = P\left[Z_i^{(v)} \leq z_i^{(v)} \mid H_{i+1} > 0 \cap Z_i^{(v)} > 0\right] \qquad (3.20)$$

(2) the conditional distribution of H_{i+1} given $Z_i^{(v)} = z_i^{(v)}$, with $z_i^{(v)} \geq 0$, consists of two parts:

$$F_{H_{i+1}|Z_i^{(v)}}\left(h_{i+1} \mid Z_i^{(v)} = 0\right) = P\left[H_{i+1} \leq h_{i+1} \mid Z_i^{(v)} = 0\right] =$$

$$\frac{(1 - p_{10} - p_{01} - p_{11}) + p_{10} \cdot F_{H_{i+1},0}^{(+,0)}(h_{i+1})}{1 - p_{01} - p_{11}} \qquad (3.21)$$

and

$$F_{H_{i+1}|Z_i^{(v)}}\left(h_{i+1} \mid Z_i^{(v)} = z_i^{(v)} \cap Z_i^{(v)} > 0\right) = P\left[H_{i+1} \leq h_{i+1} \mid Z_i^{(v)} = z_i^{(v)} \cap Z_i^{(v)} > 0\right] =$$

$$\frac{p_{01} \cdot f_{0,Z_i^{(v)}}^{(0,+)}\left(z_i^{(v)}\right) + p_{11} \cdot f_{Z_i^{(v)}}^{(+,+)}\left(z_i^{(v)}\right) \cdot F_{H_{i+1}|Z_i^{(v)}}^{(+,+)}\left(h_{i+1} \mid z_i^{(v)}\right)}{p_{01} \cdot f_{0,Z_i^{(v)}}^{(0,+)}\left(z_i^{(v)}\right) + p_{11} \cdot f_{Z_i^{(v)}}^{(+,+)}\left(z_i^{(v)}\right)} \qquad (3.22)$$

where $f_{Z_i^{(v)}}^{(+,+)}\left(z_i^{(v)}\right)$ is the density function related to $F_{Z_i^{(v)}}^{(+,+)}\left(z_i^{(v)}\right)$ and $F_{H_{i+1}|Z_i^{(v)}}^{(+,+)}\left(h_{i+1} \mid z_i^{(v)}\right)$ is the conditional distribution of H_{i+1}, given $H_{i+1} > 0$ and $Z_i^{(v)} = z_i^{(v)}$ when $z_i^{(v)} \geq 0$.

Analogous formulations may be derived for $F_{Z_i^{(v)}|H_{i+1}}\left(z_i^{(v)} \mid h_{i+1}\right)$.

Similar to the ARMA models, this methodology is valid for stationary processes only. In order to overcome the problem of non-stationarity, rainfall data must be grouped by months or seasons. The model parameters can then be estimated for a given month or season. According to the definition in Eq. (3.13), the construction of $F_{H_{i+1},Z_i^{(v)}}\left(h_{i+1},z_i^{(v)}\right)$ requires:

(1) the individuation of the process 'memory' extension v (Section 3.2.1.1);

(2) the optimal choice of the function depending on v antecedent rainfall heights defining $Z_i^{(v)}$ (Section 3.2.1.2);

(3) evaluation of p_{10}, p_{01}, p_{11}, $F_{H_{i+1},0}^{(+,0)}\left(h_{i+1}\right)$, $F_{0,Z_i^{(v)}}^{(0,+)}\left(z_i^{(v)}\right)$, $F_{H_{i+1},Z_i^{(v)}}^{(+,+)}\left(h_{i+1},z_i^{(v)}\right)$ (Section 3.2.1.3).

In Section 3.2.1.4, the methodology for rainfall nowcasting is described. In Section 3.2.1.5, an example of a developed model is reported.

3.2.1.1. Extension of Memory v

The linear stochastic dependence between the random variables H_{i+1} and a generic H_{i-j}, where $j = 0,1,2,...$, is considered negligible if the corresponding sample coefficient of partial autocorrelation is less than a given value close to zero.

The results obtained with this approach are similar to the more rigorous and less simple general method (Sirangelo et al. 2007). Using the general procedure, the absence of a linear stochastic dependence should be tested verifying that the sample coefficient of partial autocorrelation, between such random variables, exhibits a value inside a confidence interval.

This condition does not permit the rejection of the hypothesis concerning null value for the corresponding theoretical quantity.

In order to evaluate the extension of the 'memory', the absence of significant partial autocorrelation must be checked for increasing values of v. For each value of v, the negligibility must be verified for all the sample coefficients of partial autocorrelation characterized by a couple of primary subscripts H_{i+1}, H_{i-v}; H_{i+1}, H_{i-v-1}; ... and by secondary subscripts

$H_i, H_{i-1}, \ldots, H_{i-v+1}$. In other words, partial autocorrelation must be evaluated in the following way:

$$\rho_{H_{i+1}H_{i-v+1-m} \cdot H_i \ldots H_{i-v+1}} = \frac{\widetilde{\lambda}_{1(v+2)}^{(v+2)}}{\sqrt{\widetilde{\lambda}_{11}^{(v+2)}\widetilde{\lambda}_{(v+2)(v+2)}^{(v+2)}}} \qquad m = 1,2,\ldots \qquad (3.23)$$

where $\widetilde{\lambda}_{1(v+2)}^{(v+2)}$, $\widetilde{\lambda}_{11}^{(v+2)}$ and $\widetilde{\lambda}_{(v+2)(v+2)}^{(v+2)}$ are the cofactors of the Laurent edged matrix:

$$\left[\Lambda^{(v+2)}\right] = \begin{bmatrix} 1 & \rho_1 & \rho_2 & \cdots & \rho_v & \rho_{v+m} \\ \rho_1 & 1 & \rho_1 & \cdots & \rho_{v-1} & \rho_{v+m-1} \\ \rho_2 & \rho_1 & 1 & \cdots & \rho_{v-2} & \rho_{v+m-2} \\ \cdots & \cdots & \cdots & \cdots & \cdots & \cdots \\ \rho_v & \rho_{v-1} & \rho_{v-2} & \cdots & 1 & \rho_1 \\ \rho_{v+m} & \rho_{v+m-1} & \rho_{v+m-2} & \cdots & \rho_1 & 1 \end{bmatrix} \qquad (3.24)$$

in which the autocorrelation $\rho_{H_{i+1}H_{i-k+1}}$, depending only on lag k for the hypothesis of stationary stochastic process, are indicated as ρ_k.

Practically, evaluation of $\rho_{H_{i+1}H_{i-v+1-m} \cdot H_i \ldots H_{i-v+1}}$ must be performed using the sample coefficients of partial autocorrelation $r_{H_{i+1}H_{i-v+1-m} \cdot H_i \ldots H_{i-v+1}}$, $m = 1,2,\ldots$, obtained by an observed sample $h_1, h_2, \ldots, h_N$ of rainfall depths cumulated over time intervals of duration Δt, after estimation of ρ_k by sample autocorrelation coefficients r_k.

Finally, the extension of the 'memory' can be estimated by using the sample maximum absolute scattering technique, as follows:

$$\chi_r(v) = \max_{1 \leq m < \infty} \left| r_{H_{i+1}H_{i-v+1-m} \cdot H_i \ldots H_{i-v+1}} \right| \qquad m = 1,2,\ldots \qquad (3.25)$$

The extension of the process 'memory' can be fixed equal to the minimum value of v for which $\chi_r(v)$ results less than a fixed critical value $\chi_{r,cr}$.

3.2.1.2. Structure of the Random Variable Z

The criterion to define a functional dependence between the random variable $Z_i^{(v)}$ and the random variables $H_i, H_{i-1}, ..., H_{i-v+1}$ can be constituted by the maximization of the coefficient of linear correlation $\rho_{H_{i+1}Z_i^{(v)}}$ between the same $Z_i^{(v)}$ and the random variable H_{i+1}. This choice allows the best prediction of rainfall heights H_{i+1} during a storm event, after the joint probability density (Eq. 3.6) is identified. Let $Z_i^{(v)} = g\left(H_i, H_{i-1}, ..., H_{i-v+1}\right)$ be the function that satisfies the criterion stated here. Consequently, all the functions defined as:

$$g^*\left(H_i, H_{i-1}, ..., H_{i-v+1}\right) = a + b\, g\left(H_i, H_{i-1}, ..., H_{i-v+1}\right) \qquad (3.26)$$

with a and b generic constants, satisfy the same criterion, i.e. the criterion of maximization of the coefficient of linear correlation between $Z_i^{(v)}$ and H_{i+1} identifies a 'class of functions', defined by Eq. (3.26).

Clearly, the more suitable choice for $g(.)$ into Eq.(3.26) is the linear function:

$$Z_i^{(v)} = \beta + \sum_{j=0}^{v-1} \alpha_j' H_{i-j} \qquad (3.27)$$

in which the coefficients of the linear relationship that maximize $\rho_{H_{i+1}Z_i^{(v)}}$ are simply the coefficients of the linear partial regression among H_{i+1} and $H_i, H_{i-1}, ..., H_{i-v+1}$:

$$E\left[H_{i+1} | H_i', H_{i-1}', ..., H_{i-v+1}'\right] = \mu_H + \sum_{j=0}^{v-1} \alpha_j' H_{i-j}' \qquad (3.28)$$

where $E[(.)]$ is the expected value operator (see Appendix A for further details), $\mu_H = E[H_n]$. The value of $E[(.)]$ is constant for every n because of the hypothesis of the stationary stochastic process and $H_i' = H_i - \mu_H, H_{i-1}' = H_{i-1} - \mu_H,\ H_{i-v+1}' = H_{i-v+1} - \mu_H$.

The $\alpha_j^{'}$ coefficients are evaluated as follows:

$$\alpha_j^{'} = -\tilde{\lambda}_{1(j+2)}^{(v+1)} \Big/ \tilde{\lambda}_{11}^{(v+1)}$$

(3.29)

where $\tilde{\lambda}_{1(j+2)}^{(v+1)}$ e $\tilde{\lambda}_{11}^{(v+1)}$ are the cofactors of the Laurent matrix:

$$\left[\Lambda^{(v+1)}\right] = \begin{bmatrix} 1 & \rho_1 & \rho_2 & \cdots & \rho_v \\ \rho_1 & 1 & \rho_1 & \cdots & \rho_{v-1} \\ \rho_2 & \rho_1 & 1 & \cdots & \rho_{v-2} \\ \cdots & \cdots & \cdots & \cdots & \cdots \\ \rho_v & \rho_{v-1} & \rho_{v-2} & \cdots & 1 \end{bmatrix}$$

(3.30)

These coefficients can be readily evaluated by considering an observed sample $h_1, h_2, \ldots, h_N$ of rainfall heights aggregated over time intervals of duration Δt. Moreover, as a consequence of the non-negative character of the involved random variables, it should be highlighted that the coefficients $\alpha_j^{'}$ must satisfy the conditions $\alpha_j^{'} \geq 0$ for $j = 0,1,\ldots,v-1$.

The variable $Z_i^{(v)}$ can then be defined as:

$$Z_{*i}^{(v)} = \mu_H + \sum_{j=0}^{v-1} \alpha_j^{'} H_{i-j}^{'}$$

(3.31)

which is equal to the conditional expected value $E\left[H_{i+1} | H_i^{'}, H_{i-1}^{'}, \ldots, H_{i-v+1}^{'}\right]$. Rewriting Eq. (3.31) by using $H_i, H_{i-1}, \ldots, H_{i-v+1}$, the equation becomes:

$$Z_{*i}^{(v)} = \sum_{j=0}^{v-1} \alpha_j^{'} H_{i-j} + \left(1 - \sum_{j=0}^{v-1} \alpha_j^{'}\right) \mu_H$$

(3.32)

Introducing the standardized coefficients α_j :

$$\alpha_j = \alpha'_j \bigg/ \sum_{\kappa=0}^{\nu-1} \alpha'_\kappa \tag{3.33}$$

for which the conditions $0 < \alpha_j \leq 1$, for $j = 0,1,...,\nu-1$, and $\sum_{j=0}^{\nu-1} \alpha_j = 1$ apply,

and substituting Eq. (3.33) into Eq. (3.32), the expression for $Z_{*i}^{(\nu)}$ becomes:

$$Z_{*i}^{(\nu)} = \sum_{k=0}^{\nu-1} \alpha'_k \sum_{j=0}^{\nu-1} \alpha_j H_{i-j} + \left(1 - \sum_{j=0}^{\nu-1} \alpha'_j\right)\mu_H \tag{3.34}$$

Further, $Z_i^{(\nu)}$ can be expressed as:

$$Z_i^{(\nu)} = \sum_{j=0}^{\nu-1} \alpha_j H_{i-j} \tag{3.35}$$

The variables $Z_{*i}^{(\nu)}$ and $Z_i^{(\nu)}$ can be shown to be connected by a linear transformation, i.e. they belong to the class of function, defined by Eq. (3.26). Consequently, the correlation coefficient $\rho_{H_{i+1}, Z_i^{(\nu)}}$ is equal to $\rho_{H_{i+1}, Z_{*i}^{(\nu)}}$. Equation (3.35) shows that the random variable $Z_i^{(\nu)}$ can be regarded as a weighted average of the ν antecedent rainfall heights with weights expressed by the coefficients α_j.

3.2.1.3. Evaluation of p_{10}, p_{01}, p_{11}, $F_{H_{i+1},0}^{(+,0)}(h_{i+1})$, $F_{0,Z_i^{(\nu)}}^{(0,+)}(z_i^{(\nu)})$, $F_{H_{i+1},Z_i^{(\nu)}}^{(+,+)}(h_{i+1}, z_i^{(\nu)})$

On the basis of the joint sample of data $\{(h_{i+1}, z_i^{(\nu)}), i=1,...,N\}$, the probabilities p_{10}, p_{01}, p_{11} may be estimated by the frequencies N_{10}/N, N_{01}/N and N_{11}/N of the events $H_{i+1} > 0 \cap Z_i^{(\nu)} = 0$, $H_{i+1} = 0 \cap Z_i^{(\nu)} > 0$ and $H_{i+1} > 0 \cap Z_i^{(\nu)} > 0$.

For the univariate functions $F_{H_{i+1},0}^{(+,0)}(h_{i+1}), F_{0,Z_i^{(v)}}^{(0,+)}(z_i^{(v)})$, several distributions may be tested. For example, Gamma (Kottegoda and Rosso 1997), Weibull (Kottegoda and Rosso 1997), Pareto (Kleiber and Kotz 2003) and Log-Logistic (Kleiber and Kotz 2003) families (see Appendix A for details).

Regarding the structure of $F_{H_{i+1},Z_i^{(v)}}^{(+,+)}(h_{i+1},z_i^{(v)})$, see Sections 3.2.3-3.2.5 for the suggested procedures.

3.2.1.4. Methodology for Rainfall Nowcasting

Figure 3.1 shows a schematization on the use of this modelling technique for rainfall nowcasting.

(1) Let t_0 be the forecast instant and $t_1...t_N$ be the successive time instants in which a rainfall forecast is required. The period $(t_N - t_0)$ is the forecasting lead time.

(2) At t_0, $Z_i^{(v)}$ variable is a linear function of v observed rainfall data. For the forecast at time t_1, substitute the $Z_i^{(v)}$ value into Eq. (3.21) or Eq. (3.22), using the Monte Carlo technique (Kroese et al. 2011) it is possible to generate 1,000-10,000 values of H_{i+1}. In detail, the generation is made by the formula $h_{i+1} = F_{H_{i+1}|Z_i^{(v)}}^{-1}(R_U \mid Z_i^{(v)} = z_i^{(v)})$, where $R_U \in [0;1]$ is a generated random number. Numerically, the formula is solved by using bracketing and regula falsi techniques or other similar techniques (Press et al. 1988).

(3) At t_2, each generation carried out at t_1 is taken into account. In this case, another 1,000-10,000 $Z_i^{(v)}$ values are calculated, each composed by $(v-1)$ observed rainfall data and the generated value of rainfall height in the previous step. For each $Z_i^{(v)}$ value, the corresponding H_{i+1} value is obtained, by using the same procedure described in Step (2).

(4) Step (3) is repeated up until time t_N. For each generation of H_{i+1} and t_i, with $i > 2$, $Z_i^{(v)}$ value is a linear function of $(v - i + 1)$ observed rainfall data and $(i-1)$ simulated data. Further, the temporal extension of the rainfall forecast period $(t_N - t_0)$ should not exceed the memory period, $v\,\Delta t$. This is because beyond this limit the results

become unconditional on measured data. As such, updating with new observed rainfall heights in real time is then required.

Figure 3.1 shows the results from Step (1)-(4), which include the possible realizations in the forecasting lead time. This is called a "spaghetti" plot.

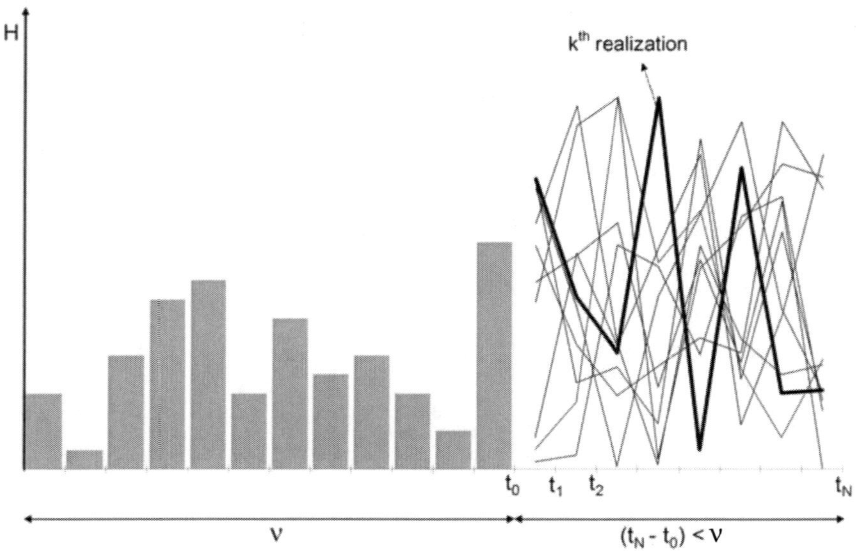

Figure 3.1. Qualitative example of a "spaghetti" plot.

The Monte Carlo technique is adopted because of the complexity of determining analytical probabilistic distributions for rainfall heights relative to forecasting time instants successive to the first one. In fact, for these distributions, convolution operations should be required. After carrying out the procedure described in Steps (2)-(4), for each time instant a PQPF may be provided, for example in terms of assigned percentiles with a plot similar to Figure 2.3.

3.2.1.5. Application of Methodology for Rainfall Nowcasting

As an example, a model named Prediction of Rainfall Inside Storm Events or PRAISE (Sirangelo et al. 2007) was developed at the Department of Soil Conservation, University of Calabria, Italy, and applied in cooperation with the Italian National Department of Civil Protection and Regional Administrations. PRAISE was calibrated by applying the split-sample approach to the hourly data of rain gauge network in the Calabria region in southern Italy (Figure 3.2). The rainfall heights used for calibration were the observed data during the rainy season of October 1 through May 31 for the years 1990-2005. In this rainy season, the correlation structure, mean and variance of the sample appear significantly homogeneous (De Luca 2006).

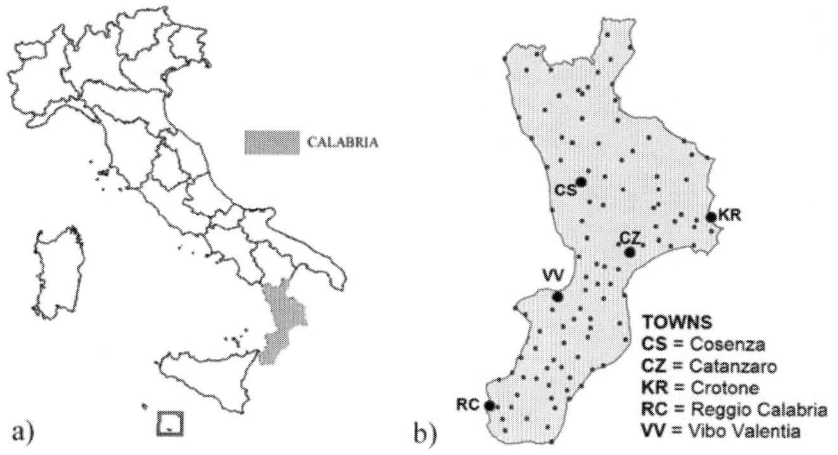

Figure 3.2.a) Location of Calabria region; b) Rain gauges network.

Concerning model validation, Figure 3.3 shows the output of PRAISE Model for Montalto Uffugo rain gauge, near to Cosenza, for the event of December 12-13, 2008. A linear function was adopted for the $Z_i^{(v)}$ variable, and a memory $v = 8$ hours was obtained by considering $\chi_{r,cr} = 0.025$. The Weibull distributions were used for $F_{H_{i+1},0}^{(+,0)}(h_{i+1}), F_{0,Z_i^{(v)}}^{(0,+)}(z_i^{(v)})$, and power transformation of the bivariate Moran-Downton exponential distribution was adopted for $F_{H_{i+1},Z_i^{(v)}}^{(+,+)}(h_{i+1}, z_i^{(v)})$ (see Section 3.2.3 for further details).

For the selected event, the observed rainfall on December 13, 2008 was 175.8 mm, which corresponds to a return period of about 180 years. In the application, the memory extension of observed rainfall was extended from 18:00 LT (Local Time) on December 12th to 2:00 LT on the 13th, the forecasting period ranges from 2:00 LT to 8:00 LT on the 13th. The observed rainfall was compared with the 80%, 90% and 95% percentiles of the probabilistic distributions for every forecasting hour. The comparison emphasizes the ability of PRAISE to identify confidence intervals, which

include the observed rainfall heights. See Appendix B for further details about
the meaning of confidence intervals.

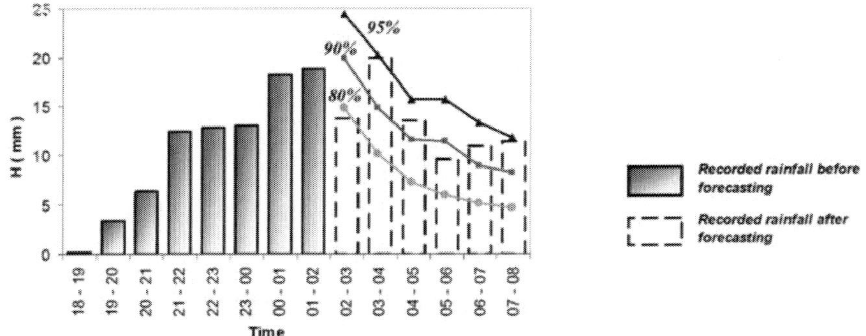

Figure 3.3. Montalto Uffugo rain gauge: PRAISE application for December 13, 2008.

3.2.2. Spatiotemporal Models

With reference to the temporal models as described in Section 3.2.1,
spatiotemporal models can be developed by adopting the same approach that
was used for the STARMA models (Section 3.1). In general, a multiple time
series should be considered and the dimensions of the multivariate analysis
correspond to the number of sites (rain gauges or cells) to be analyzed.

A simplification can be done by assuming a local spatial dependence, i.e.
an assigned point (cell) is influenced only by its neighboring sites.
Consequently, a local spatiotemporal model can be realized by defining, for
each site, the following variables (Figure 3.4):

(1) H_{i+1} , corresponding to the rainfall height on the forecasting interval
     $[i\Delta t;(i+1)\Delta t]$ related to the cell having an area $\Delta x \Delta y$;

(2) with the same approach as in Section 3.2.1, for a given lag in time
     $v\Delta t$, and for a given pixel where the future rainfall H_{i+1} must be
     estimated, $Z_i^{(v)}$ is a (linear or nonlinear) function of v antecedent
     precipitation;

(3) W_{i+1} , corresponding to a weighted average of the simultaneous
     rainfall heights over the four closest neighboring pixels:

$$W_{i+1} = \sum_{j=1}^{4} \beta_j' H_{i+1}^{(j)} \tag{3.36}$$

It is apparent that larger neighborhoods or a different weighting procedure may be used, allowing for influence from a larger region or anisotropy. In either case, however, a greater number of parameters are required. The preceding model type may be applied to a regular (e.g. radar data) or an irregular spatial domain.

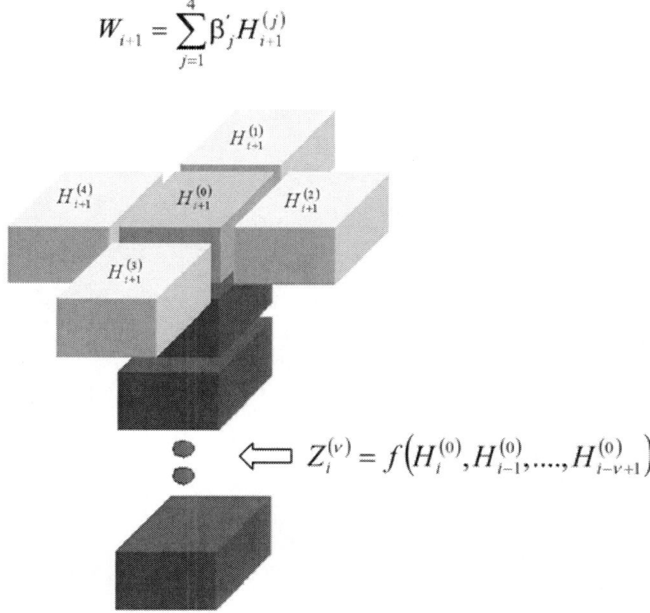

Figure 3.4. Definition of random variable for each pixel (adapted from Versace et al. 2009).

In the latter case, for example starting from a rain gauge network, it is possible to carry out an interpolation of historical rainfall data on a regular discretized domain (according to the rain gauge network density), by using, for example, the surface spline technique (Yu 2001).

Moreover, in the case of a high-resolution timestep Δt, such as hourly or sub-hourly, storm advection can be reproduced, in a stochastic way, by the correlation structure referred to the spatial neighborhood approach. In the following, for simplicity, the notations for superscripts and subscripts related to the random variables are removed.

With the same procedure as described in Section 3.2.1, a trivariate joint probability density function is defined for each pixel as:

$$
f_{H,W,Z}(h,w,z) = p_{000}\delta(h)\cdot\delta(w)\cdot\delta(z) + p_{100}\cdot f_{H,0,0}^{(+,0,0)}(h)\cdot\delta(w)\cdot\delta(z) +
$$

$$
p_{010}\cdot f_{0,W,0}^{(0,+,0)}(w)\cdot\delta(h)\cdot\delta(z) + p_{001}\cdot f_{0,0,Z}^{(0,0,+)}(z)\cdot\delta(h)\cdot\delta(w) +
$$

$$
p_{110}\cdot f_{H,W,0}^{(+,+,0)}(h,w)\cdot\delta(z) + p_{101}\cdot f_{H,0,Z}^{(+,0,+)}(h,z)\cdot\delta(w) +
$$

$$
p_{011}\cdot f_{0,W,Z}^{(0,+,+)}(w,z)\cdot\delta(h) + p_{111}\cdot f_{H,W,Z}^{(+,+,+)}(h,w,z)
$$

$$(3.37)$$

in which:

$$
p_{000} = P[H = 0 \cap W = 0 \cap Z = 0] \tag{3.38}
$$

$$
p_{100} = P[H > 0 \cap W = 0 \cap Z = 0] \tag{3.39}
$$

$$
p_{010} = P[H = 0 \cap W > 0 \cap Z = 0] \tag{3.40}
$$

$$
p_{001} = P[H = 0 \cap W = 0 \cap Z > 0] \tag{3.41}
$$

$$
p_{110} = P[H > 0 \cap W > 0 \cap Z = 0] \tag{3.42}
$$

$$
p_{101} = P[H > 0 \cap W = 0 \cap Z > 0] \tag{3.43}
$$

$$
p_{011} = P[H = 0 \cap W > 0 \cap Z > 0] \tag{3.44}
$$

$$
f_{H,0,0}^{(+,0,0)}(h)\cdot dh = P[h \le H < h + dh \,|\, H > 0 \cap W = 0 \cap Z = 0] \tag{3.45}
$$

$$
f_{0,W,0}^{(0,+,0)}(w)\cdot dw = P[w \le W < w + dw \,|\, H = 0 \cap W > 0 \cap Z = 0] \tag{3.46}
$$

$$f_{0,0,Z}^{(0,0,+)}(z) \cdot dz = P\big[z \leq Z < z+dz \big| H=0 \cap W=0 \cap Z>0\big] \tag{3.47}$$

$$f_{H,W,0}^{(+,+,0)}(h,w) \cdot dh \cdot dw =$$

$$P\big[h \leq H < h+dh \cap w \leq W < w+dw \big| H>0 \cap W>0 \cap Z=0\big] \tag{3.48}$$

$$f_{H,0,Z}^{(+,0,+)}(h,z) \cdot dh \cdot dz =$$

$$P\big[h \leq H < h+dh \cap z \leq Z < z+dz \big| H>0 \cap W=0 \cap Z>0\big] \tag{3.49}$$

$$f_{0,W,Z}^{(0,+,+)}(w,z) \cdot dw \cdot dz =$$

$$P\big[w \leq W < w+dw \cap z \leq Z < z+dz \big| H=0 \cap W>0 \cap Z>0\big] \tag{3.50}$$

$$f_{H,W,Z}^{(+,+,+)}(h,w,z) \cdot dh \cdot dw \cdot dz =$$

$$P\big[h \leq H < h+dh \cap w \leq W < w+dw \cap z \leq Z < z+dz \big| H>0 \cap W>0 \cap Z>0\big]$$
$$\tag{3.51}$$

$\delta(\cdot)$ is Dirac's function and $p_{000} + p_{100} + p_{010} + p_{001} + p_{110} + {}$
$+ p_{101} + p_{001} + p_{111} = 1$.

The conditional distribution $f_{H|W,Z}(h|w,z)$, necessary to perform the forecast, is characterized by four separate cases, due to null or positive values for both W and Z random variables.

Considering the cumulative distribution function (CDF) $F_{H|W,Z}(h|w,z)$, the following expressions are obtained:

- if $W=0$ and $Z=0$ then

$$F_{H|W,Z}(h|w,z) = \frac{p_{000} + p_{100} F_{H,0,0}^{(+,0,0)}(h)}{p_{000} + p_{100}} \tag{3.52}$$

where $F_{H,0,0}^{(+,0,0)}(h)$ is the CDF of $f_{H,0,0}^{(+,0,0)}(h)$ and $\dfrac{p_{000}}{p_{000}+p_{100}}$ is the probability referred to the event $H=0\,|\,W=0\cap Z=0$;

- if $W>0$ and $Z=0$ then

$$F_{H|W,Z}(h|w,z)=\frac{p_{010}\cdot f_{0,W,0}^{(0,+,0)}(w)+p_{110}\cdot F_{H|W,0}^{(+,+,0)}(h|w)\cdot f_{W}^{(+,+,0)}(w)}{p_{010}\cdot f_{0,W,0}^{(0,+,0)}(w)+p_{110}\cdot f_{W}^{(+,+,0)}(w)} \qquad (3.53)$$

where $F_{H|W,0}^{(+,+,0)}(h|w)$ is the CDF of the conditional density $f_{H|W,0}^{(+,+,0)}(h|w)$, $f_{W}^{(+,+,0)}(w)$ is the marginal density with the respect to W, derived from $f_{H,W,0}^{(+,+,0)}(h,w)$, and $\dfrac{p_{010}\cdot f_{0,W,0}^{(0,+,0)}(w)}{p_{010}\cdot f_{0,W,0}^{(0,+,0)}(w)+p_{110}\cdot f_{W}^{(+,+,0)}(w)}$ is the probability referred to the event $H=0\,|\,W>0\cap Z=0$;

- if $W=0$ and $Z>0$ then

$$F_{H|W,Z}(h|w,z)=\frac{p_{001}\cdot f_{0,0,Z}^{(0,0,+)}(z)+p_{101}\cdot F_{H|0,Z}^{(+,0,+)}(h|z)\cdot f_{Z}^{(+,0,+)}(z)}{p_{001}\cdot f_{0,0,Z}^{(0,0,+)}(z)+p_{101}\cdot f_{Z}^{(+,0,+)}(z)} \qquad (3.54)$$

where $F_{H|0,Z}^{(+,0,+)}(h|z)$ is the CDF of the conditional density $f_{H|0,Z}^{(+,0,+)}(h|z)$, $f_{Z}^{(+,0,+)}(z)$ is the marginal density with respect to Z, derived from $f_{H,0,Z}^{(+,0,+)}(h,z)$, and $\dfrac{p_{001}\cdot f_{0,0,Z}^{(0,0,+)}(z)}{p_{001}\cdot f_{0,0,Z}^{(0,0,+)}(z)+p_{101}\cdot f_{Z}^{(+,0,+)}(z)}$ is the probability referred to the event $H=0\,|\,W=0\cap Z>0$;

- if $W>0$ and $Z>0$ then

$$F_{H|W,Z}(h|w,z)=\frac{p_{011}\cdot f_{0,W,Z}^{(0,+,+)}(w,z)+p_{111}\cdot F_{H|W,Z}^{(+,+,+)}(h|w,z)\cdot f_{W,Z}^{(+,+,+)}(w,z)}{p_{011}\cdot f_{0,W,Z}^{(0,+,+)}(w,z)+p_{111}\cdot f_{W,Z}^{(+,+,+)}(w,z)}$$

$$(3.55)$$

where $F_{H|W,Z}^{(+,+,+)}(h|w,z)$ is the CDF of the conditional density $f_{H|W,Z}^{(+,+,+)}(h|w,z)$, $f_{W,Z}^{(+,+,+)}(w,z)$ is the marginal density with the respect to W and Z, derived from $f_{H,W,Z}^{(+,+,+)}(h,w,z)$, and

$$\frac{p_{011} \cdot f_{0,W,Z}^{(0,+,+)}(w,z)}{p_{011} \cdot f_{0,W,Z}^{(0,+,+)}(w,z) + p_{111} \cdot f_{W,Z}^{(+,+,+)}(w,z)}$$ is the probability referred to the

event $H = 0 \,|\, W > 0 \cap Z > 0$.

Similar to Section 3.2.1, the construction of $f_{H,W,Z}(h,w,z)$ for each cell requires:

(1) the individuation of the process 'memory' extension v (Section 3.2.2.1);

(2) the optimal choice of the function depending on v antecedent rainfall heights defining Z (Section 3.2.2.1);

(3) estimation of coefficients $\beta_j, j = 1,...,4$ (Section 3.2.2.2);

(4) evaluation of all the probabilities and CDFs from Eq. (3.37) (Section 3.2.2.3).

In Section 3.2.2.4, the methodology for rainfall nowcasting is described. In Section 3.2.1.5, an example of the developed model is reported.

3.2.2.1. Extension of Memory v and Definition of Random Variable Z

Similar to Section 3.2.1.1, the extension of the temporal 'memory' v, for every cell, can be assumed equal to the minimum value of v for which the sample maximum absolute scattering $\chi_r(v)$ (Sirangelo et a., 2007 and Eq. 3.25) is less than a fixed critical value $\chi_{r,cr}$.

Regarding the definition of random variable Z for each cell, if a linear function of antecedent v rainfall heights is adopted (Eq.3.35), the coefficients α_j can be estimated by maximization of the coefficient of linear correlation ρ_{HZ} (Section 3.2.1.2). If v is large, then a great number of parameters must be estimated. Thus, a technique of linear filtering may be required, allowing that the coefficients α_j depend on a reduce number of parameters. As an example, Versace et al. (2009) proposed the gamma-power function as a filter (De Luca,

2006), which depends on three parameters and can provide a good fitting to the estimated α_j. Thus, α_j can be calculated as:

$$\alpha_j = \frac{P\left[a,\left((j+1)b\,\Delta t\right)^c\right] - P\left[a,\left(jb\,\Delta t\right)^c\right]}{P\left[a,\left(vb\,\Delta t\right)^c\right]} \tag{3.56}$$

where $P(a,x) = \dfrac{1}{\Gamma(a)}\displaystyle\int_0^x e^{-t}t^{a-1}dt$ is the incomplete gamma function (Abramowitz and Stegun 1970).

The values of a, b, c are to be evaluated as follows:

(1) for every set *(a, b, c)*, the sample variable $z_i^{(v)}$ for $i = v, v+1, ..., N-1$, can be evaluated;

(2) the optimal $\hat{a}$, $\hat{b}$, $\hat{c}$ can then be obtained by maximizing the sample linear correlation coefficient, r_{HZ}, between H and Z. The numerical procedure (Press et al. 1988) to maximize r_{HZ} must be adopted.

3.2.2.2. Estimation of Coefficients β_j

With regard to the random variable W, for each cell, the coefficients β_j can be evaluated as:

$$\beta_j = \frac{\rho'_{j,0}}{\displaystyle\sum_{j=1}^{4}\rho'_{j,0}} \tag{3.57}$$

where $j = 1,2,3,4$, and $\rho'_{1,0}, \rho'_{2,0}, \rho'_{3,0}, \rho'_{4,0}$ indicate the linear correlation coefficients between the reference cell *0* and the neighbor cells *1, 2, 3* and *4* (Figure 3.4). In the case in which the schematization of regular cells should derive from a rain gauge network, the estimation of β_j can be carried out as follows:

(1) The spatial domain is discretized by using a resolution according to the network density. For example, if the mean distance among the rain gauges is 10 km, then this distance is used as the spatial resolution of the domain.

(2) The sample directional spatial correlograms can be analyzed using simultaneous rainfall heights. On the horizontal axis, only distances between rain gauges over a range [5 km; 15 km], representing spatial resolution of the region, should be considered. As an example, for the Calabria region, Versace et al. (2009) discretized the domain with a resolution of 10 km (see Section 3.2.2.5 for further details); the range considered was [5 km; 15 km]. Four circular sectors of 90 degrees, centered in the NE, SE, SW, and NW directions, can be adopted to discriminate the class of directions. Figure 3.5 shows the results obtained for the Calabria region (Versace et al. 2009), in which sample correlation values appear to be independent on the directions. Thus, the rain fields can be considered as local isotropic, and $\beta_j' = 1/4$, $j = 1, 2, 3, 4$.

3.2.2.3. Evaluation of Probabilities and CDFs

Similar to Section 3.2.1.3, the probabilities p_{111}, p_{110}, p_{101}, p_{011}, p_{100}, p_{010} and p_{001} can be estimated by the frequencies of the corresponding events. Distributions such as Gamma (Kottegoda and Rosso 1997), Weibull (Kottegoda and Rosso 1997), Pareto (Kleiber and Kotz 2003) and Log-Logistic (Kleiber and Kotz 2003) can be used for all the univariate functions (see Appendix A for further details). In Sections 3.2.3-3.2.5, bivariate and trivariate distributions are described.

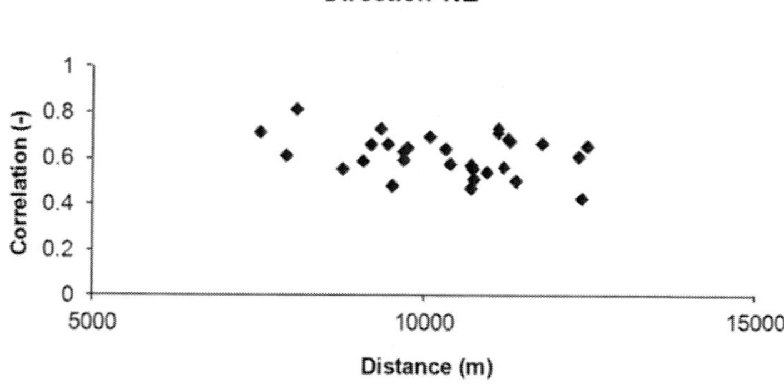

Figure 3.5. (Continued)

                     Davide Luciano De Luca

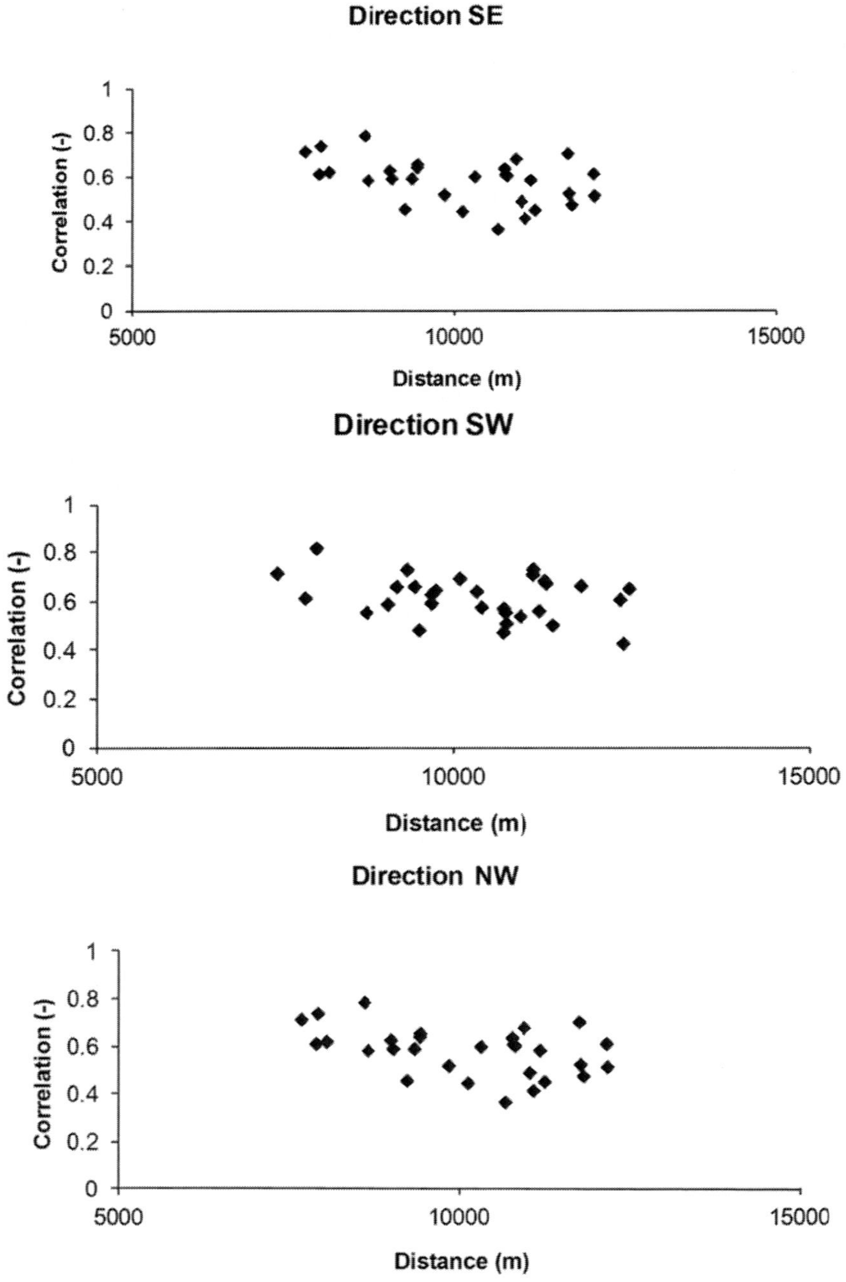

Figure 3.5. Sample directional correlograms obtained for Calabria region (from Versace et al., 2009).

3.2.2.4. Methodology for Rainfall Nowcasting

With regard to the nowcasting procedure, the generation of the rainfall heights on the entire domain differs from that of the standard Monte Carlo approach. In fact, at the current time before nowcasting, only the values of the random variable Z (as combination of observed and generated data) in each cell are known.

The values of W and H on the entire domain must be generated simultaneously. Such generations cannot be carried out independently cell-by-cell, because the variables are linked by congruence equations (Eq. 3.36). This problem has been solved using the following "Chess-Board" algorithm (Figure 3.6), proposed by Versace et al. (2009):

(1) Starting from the "0" cells of the spatial domain, the value of Z is known. The generation is then made by using the random number $R_U^{(0)}$, and the formula $h^{(0)} = F_{H|Z}^{-1}\left(R_U^{(0)}|W^{(0)} \geq 0, Z^{(0)} = z^{(0)}\right)$, i.e. the variable H is generated supposing zero as the lower bound for W. This type of generation can be adopted because rainfall heights in the "0" cells are supposed to be independent of each other.

(2) With regard to the "1" cells, knowing the value of Z, $W^{(1)}$ is set equal to the linear combination of the $H^{(0)}$ in the neighboring "0" cells. Consequently, generation is made by the formula $h^{(1)} = F_{H|W,Z}^{-1}\left(R_U^{(1)}|W^{(1)} = w^{(1)}, Z^{(1)} = z^{(1)}\right)$ using the random number $R_U^{(1)}$.

This procedure may be adopted for each forecast instant, starting from t_0, similar to the temporal model in Figure 3.1. In this case, for each cell and t_i, with $i > 2$, the generation of Z value is a combination of $(\nu - i + 1)$ observed rainfall data and $(i-1)$ simulated data. Consequently, it is clear that the temporal extension of the forecast $(t_N - t_0)$ should not exceed the memory $\nu \Delta t$ Beyond this limit the results become unconditional on measured data, and then a model update by observed precipitation is necessary.

3.2.2.5. Application of Methodology for Rainfall Nowcasting

As an example, Versace et al. (2009) developed the PRAISEST model (Prediction of Rainfall Amount Inside Storm Events: Space and Time) which constitutes the spatiotemporal extension of PRAISE (Section 3.2.1).

PRAISEST was calibrated using the hourly data from 1990-2005 of the rain gauge network of the Calabria region (Figure 3.2). Moreover:

0	1	0	1	0
1	0	1	0	1
0	1	0	1	0
1	0	1	0	1
0	1	0	1	0

Figure 3.6. Scheme of "Chess-Board" algorithm (from Versace et al. 2009).

- the region was discretized by a 10 km x 10 km cell grid, according to the rain gauge network density;
- in order to respect the hypothesis of the stationary process, only the data measured during the 'rainy season' (i.e. October 1st – May 31st) have been used (De Luca, 2006);
- a linear function was adopted for the Z variable, a memory $v\Delta t = 8$ hours and $\beta_j = 0.25, j = 1,...,4$ (i.e. the rainfall field is assumed locally isotropic) were evaluated;
- Weibull distributions were used for all the univariate distributions. Power transformation of the bivariate and trivariate Moran-Downton exponential function was adopted for all the joint distributions (see Section 3.2.3 for further details).

The obtained trivariate probability distribution function $f_{H,W,Z}(h,w,z)$ presents 42 parameters for each pixel (Versace et al. 2009). This number is not too large. In fact, adopting a split-sample calibration, if we consider that in 10 years there are 87,600 hourly data (assuming that the rainfall process is stationary in time during the whole year) related to every cell, then the d/p (data/parameters in a generic cell) ratio approximately equaling 2,050. This ratio remains high enough (about 150) even if only positive rainfall data are considered. The validation was carried out by focusing attention on the ensemble of events within the storm development and characterized by $Z_0 > 1$ mm (where the subscript "0" relates to the previous eight hours of recorded data). These are of major interest, because a purely stochastic model should provide reliable results in the case of nowcasting "inside a storm events" (see Chapter 1). Model performance was tested on reproducing, for each of the successive six hours of forecasting, the following quantities which are not considered in the parameter estimation:

a)   mean $m_{H|Z_0>1}$ and standard deviations $s_{H|Z_0>1}$;

b)   spatial correlation $r_{HW|Z_0>1}$ and autocorrelation of lag 1 $r_{1|Z_0>1}$;

c)   dry ratios $f_{00|Z_0>1}$, wet-to-dry $f_{10|Z_0>1}$, dry-to wet $f_{01|Z_0>1}$ and wet-to-wet $f_{11|Z_0>1}$ ratios for two subsequent images (it means, for two consecutive rainfall fields in time, the frequency of the cells characterized by null rainfall for both instants, positive and then null rainfall, null and then positive rainfall, positive rainfall for both instants, respectively).

The results are presented in Figure 3.7. The histograms represent the sample values averaged over the whole spatial domain. Moreover, for each index, the mean value and the 95% confidence interval of PRAISEST simulations are reported.

Figure 3.7 shows that the model is capable of reproducing the properties of the rainfall fields.

Moreover, as an example of the model validation, the results of the application to the January 25, 2009 event, in the Calabria region, is shown in Figure 3.8. On this date, a landslide occurred near the Rogliano municipality, along the A3 "Salerno – Reggio Calabria" motorway, and killed two people. For this event, the rain fields from 3:00 LT to 11:00 LT were used as the model memory, and the period from 11:00 LT to 17:00 LT was used for simulation. The simulation period was chosen by considering the maximum six-hour rainfall period for the whole region on this day. On the abscissa, the cells are geographically sorted from left to right, and from North to South. Following the horizontal axis, the main towns of Cosenza (CS), Crotone (KR), Catanzaro (CZ), Vibo Valentia (VV) and Reggio Calabria (RC) are arranged. In Figure 3.8, the observed rainfall heights aggregated in the successive six hours of nowcasting, and the 90% and 95% percentiles of the simulated fields, are reported for each cell. The figure shows that observed rainfall heights for all but one of the cells are smaller than the 95% percentile of forecasts. In many cases, observed values are also smaller than the 90% percentile forecasts. Therefore, the forecasts by the PRAISEST model are generally in agreement with observed data in all cases.

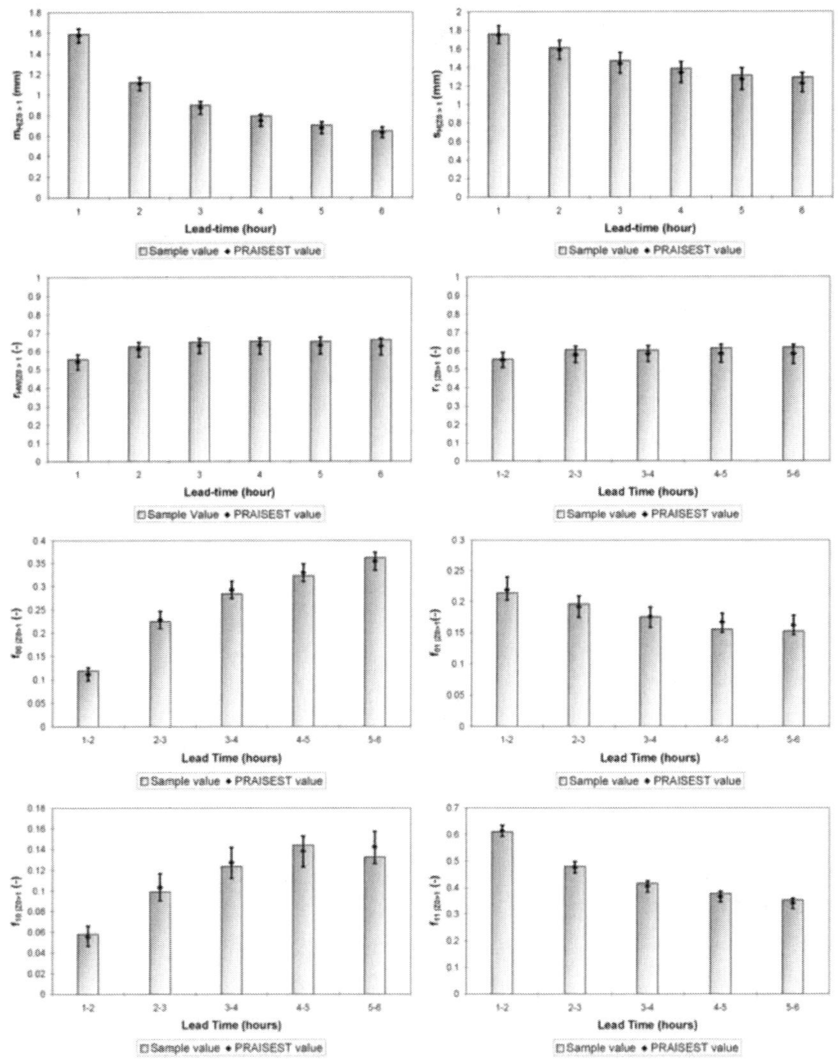

Figure 3.7. PRAISEST Model: validation results for $m_{H|Z_0>1}$, $s_{H|Z_0>1}$, $r_{HW|Z_0>1}$, $r_{1|Z_0>1}$, $f_{00|Z_0>1}$, $f_{10|Z_0>1}$, $f_{01|Z_0>1}$ and $f_{11|Z_0>1}$ (from Versace et al. 2009).

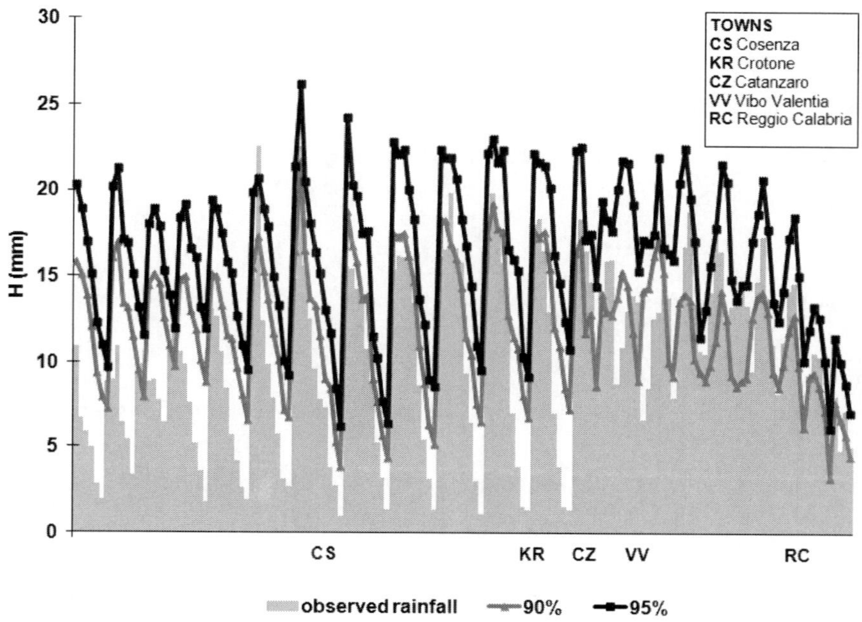

Figure 3.8. PRAISEST application for the January 25, 2009 event: comparison between observed rainfall over the interval 11:00-17:00 LT and the PRAISEST 90% and 95% percentiles forecasts.

3.2.3. The Moran-Downton Multivariate Exponential Distribution

The Moran-Downton multivariate exponential distribution (Kotz et al. 2000) for a vector $\underline{X}$ having p unit mean marginal distributions is written as:

$$f_{\underline{X}}(x) = \theta^{p-1} \cdot \exp\left(-\theta \cdot \sum_{i=1}^{p} x_i\right) \cdot S_p\left[(\theta-1)\cdot\theta^{p-1}\cdot\prod_{i=1}^{p} x_i\right] \qquad (3.58)$$

where θ is an association parameter and $S_p[q]$ is expressed as:

$$S_p[q] = \sum_{i=1}^{\infty} \frac{q^r}{(r!)^p} \qquad (3.59)$$

The Eq. (3.58) is characterized by exponential marginal density functions. The association parameter respects the condition $\theta \geq 1$ and it is related to the theoretical linear correlation coefficient between any two variables X_i and X_j, as $\rho_{i,j} = 1 - 1/\theta$. Consequently, if $\theta = 1$ then $\rho_{i,j} = 0$ and Eq. (3.58) can be written as:

$$f_{\underline{X}}(\underline{x}) = \exp\left(-\sum_{i=1}^{p} x_i\right) = \prod_{i=1}^{p} \exp(-x_i) \tag{3.60}$$

In Eq. (3.60), it is assumed that the random variables are independent of each other.

If power transformations are applied for bivariate and trivariate cases (like in PRAISE and PRAISEST models in Sections 3.2.1-3.2.2), i.e.:

$$x_1 = \lambda_h \cdot h^{\eta_h} \qquad\qquad h > 0, \lambda_h > 0, \eta_h > 0 \tag{3.61}$$

$$x_2 = \lambda_z \cdot z^{\eta_z} \qquad\qquad z > 0, \lambda_z > 0, \eta_z > 0 \tag{3.62}$$

$$x_3 = \lambda_w \cdot w^{\eta_w} \qquad\qquad w > 0, \lambda_w > 0, \eta_w > 0 \tag{3.63}$$

the marginal functions are Weibull distributions:

$$F_H(h) = 1 - \exp\left(-\lambda_h \cdot h^{\eta_h}\right) \tag{3.64}$$

$$F_Z(z) = 1 - \exp\left(-\lambda_z \cdot z^{\eta_z}\right) \tag{3.65}$$

$$F_W(w) = 1 - \exp\left(-\lambda_w \cdot w^{\eta_w}\right) \tag{3.66}$$

and the conditional density functions may be easily written. For example, $f_{H|W,Z}^{(+,+,+)}(h \mid w, z)$, related to Eq. (3.55), assumes the following expression:

$$f_{H|W,Z}^{(+,+,+)}(h \mid w,z) = \frac{f_{H,W,Z}^{(+,+,+)}(h,w,z)}{f_{W,Z}^{(+,+,+)}(w,z)} = \theta_{hwz} \cdot \lambda_h \cdot \eta_h \cdot h^{\eta_h - 1} \cdot \exp\left(-\theta_{hwz} \cdot \lambda_h \cdot h^{\eta_h}\right).$$

$$\cdot \frac{S_3\left[(\theta_{hwz} - 1) \cdot \theta_{hwz}^2 \cdot \lambda_h \cdot h^{\eta_h} \cdot \lambda_z \cdot z^{\eta_z} \cdot \lambda_w \cdot w^{\eta_w}\right]}{S_2\left[(\theta_{hwz} - 1) \cdot \theta_{hwz} \cdot \lambda_z \cdot z^{\eta_z} \cdot \lambda_w \cdot w^{\eta_w}\right]}$$

$$(3.67)$$

where the power transformation parameters (λ_h, η_h), (λ_z, η_z), (λ_w, η_w) and the association parameter θ_{hwz} are estimated from the observations $(h > 0, z > 0, w > 0)$. As θ_{hwz} is the same for each couple of random variables, it is evaluated from all of the pairs within the groups $(h > 0, z > 0)$, $(h > 0, w > 0)$ and $(w > 0, z > 0)$. The CDF $F_{H|W,Z}^{(+,+,+)}(h \mid w,z)$ will be obtained by the corresponding integral relationship (i.e. Eq. 2.3). Further, the conditional density $f_{H|W,0}^{(+,+,0)}(h \mid w)$ of Eq. (3.53) can be written as:

$$f_{H|W,0}^{(+,+,0)}(h \mid w) = \frac{f_{H,W,0}^{(+,+,0)}(h,w)}{f_W^{(+,+,0)}(w)} = \theta_{hw} \cdot \lambda_h \cdot \eta_h \cdot h^{\eta_h - 1} \cdot \exp\left(-\theta_{hw} \cdot \lambda_h \cdot h^{\eta_h}\right).$$

$$\cdot S_2\left[(\theta_{hw} - 1) \cdot \theta_{hw} \cdot \lambda_h \cdot h^{\eta_h} \cdot \lambda_w \cdot w^{\eta_w}\right]$$

$$(3.68)$$

where the power transformation parameters (λ_h, η_h), (λ_w, η_w) are different from those of Eq. (3.67), because they are evaluated from the observations $(h > 0, z = 0, w > 0)$. The same set is used for θ_{hw} estimation.

Similar expressions can be developed for $f_{H|0,Z}^{(+,0,+)}(h \mid z)$ of Eq. (3.54), for which the parameters, comprising θ_{hz}, are estimated from the record set $(h > 0, z > 0, w = 0)$:

$$f_{H|0,Z}^{(+,0,+)}(h \mid z) = \frac{f_{H,0,Z}^{(+,0,+)}(h,z)}{f_z^{(+,0,+)}(z)} = \theta_{hz} \cdot \lambda_h \cdot \eta_h \cdot h^{\eta_h - 1} \cdot \exp\left(-\theta_{hz} \cdot \lambda_h \cdot h^{\eta_h}\right).$$

$$\cdot S_2\left[(\theta_{hz} - 1) \cdot \theta_{hz} \cdot \lambda_h \cdot h^{\eta_h} \cdot \lambda_z \cdot z^{\eta_z}\right]$$

$$(3.69)$$

and for $f_{H|Z}^{(+;+)}(h \mid z)$ related to Eq.(3.22) in the PRAISE model, for which the bivariate sample $(h > 0,\ z > 0)$ must be considered:

$$f_{H|Z}^{(+,+)}(h \mid z) = \frac{f_{H,Z}^{(+,+)}(h,z)}{f_z^{(+,+)}(z)} = \theta_{hz} \cdot \lambda_h \cdot \eta_h \cdot h^{\eta_h - 1} \cdot \exp\!\left(-\theta_{hz} \cdot \lambda_h \cdot h^{\eta_h}\right) \cdot$$
$$\cdot S_2\!\left[(\theta_{hz} - 1) \cdot \theta_{hz} \cdot \lambda_h \cdot h^{\eta_h} \cdot \lambda_z \cdot z^{\eta_z}\right]$$

$$(3.70)$$

Starting from Eq. (3.58), it is easy to obtain the expression of the other probability functions. In particular, with regard to the functions $f_{H,0,0}^{(+,0,0)}(h)$, $f_{0,W,0}^{(0,+,0)}(w)$, $f_{0,0,Z}^{(0,0,+)}(z)$ of PRAISEST, and $f_{H,0}^{(+,0)}$, $f_{0,Z}^{(0,+)}(z)$ of PRAISE, if Weibull distributions are adopted, they do not require additional parameters apart from those in the power transformation. Regarding parameter estimation, for each sample combination of zero-nonzero random variables, Weibull distribution parameters may be evaluated by using the method of moments (see Appendix B for further details). For example, considering the case $(h > 0,\ z > 0,\ w > 0)$ in the PRAISEST model one obtains:

$$\hat{\eta}_h = \eta_h: \quad \Gamma\!\left(1 + 2/\eta_h\right) / \Gamma^2\!\left(1 + 1/\eta_h\right) = 1 + s_H^2 / m_H^2 \tag{3.71}$$

$$\hat{\lambda}_h = \left[\Gamma\!\left(1 + 1/\hat{\eta}_h\right) / m_H\right]^{\hat{\eta}_h} \tag{3.72}$$

$$\hat{\eta}_w = \eta_w: \quad \Gamma\!\left(1 + 2/\eta_w\right) / \Gamma^2\!\left(1 + 1/\eta_w\right) = 1 + s_W^2 / m_W^2 \tag{3.73}$$

$$\hat{\lambda}_w = \left[\Gamma\!\left(1 + 1/\hat{\eta}_w\right) / m_W\right]^{\hat{\eta}_w} \tag{3.74}$$

$$\hat{\eta}_z = \eta_z: \quad \Gamma\!\left(1 + 2/\eta_z\right) / \Gamma^2\!\left(1 + 1/\eta_z\right) = 1 + s_Z^2 / m_Z^2 \tag{3.75}$$

$$\hat{\lambda}_z = \left[\Gamma\!\left(1 + 1/\hat{\eta}_z\right) / m_Z\right]^{\hat{\eta}_z} \tag{3.76}$$

where m_H, m_W, m_Z and s_H, s_W, s_Z are, respectively, sample means and standard deviations, and $\Gamma(.)$ is the complete gamma function (Abramowitz and Stegun 1970). Eqs. (3.71), (3.73) and (3.75) must be solved using numerical methods (Press et al. 1988). Similar expressions can be developed for the other cases in the PRAISEST and PRAISE models

With regard to the association parameter, it is related to the linear correlation coefficient between any couple of nonzero random variables. For example, considering the case $(h > 0,\ z > 0, w > 0)$ in the PRAISEST model, one obtains for the couple $(h > 0,\ z > 0)$:

$$\rho_{HZ}^{(+,+,+)} = \frac{2F_1\left(-1/\eta_h,\ -1/\eta_z\ ;\ 1\ ;\ 1-1/\theta_{hwz}\right)-1}{\sqrt{\left[\dfrac{\Gamma\left(1+2/\eta_h\right)}{\Gamma^2\left(1+1/\eta_h\right)}-1\right]\cdot\left[\dfrac{\Gamma\left(1+2/\eta_z\right)}{\Gamma^2\left(1+1/\eta_z\right)}-1\right]}} \qquad (3.77)$$

where $_2F_1(a,b;c;u)$ is the hypergeometric function (Abramowitz and Stegun 1970). It must be pointed out that with higher θ_{hwz} parameter values, the $\rho_{HZ}^{(+,+,+)}$ values are also higher. Similar expressions may be developed for the other couples and cases in the PRAISEST and PRAISE models.

For the case $(h > 0,\ z > 0, w > 0)$, θ_{hwz} evaluation is performed by minimizing the following function:

$$R(\theta_{hwz}) = \ \omega_1\left(\rho_{HW}^{(+,+,+)}-r_{HW}^{(+,+,+)}\right)^2 + \omega_2\left(\rho_{HZ}^{(+,+,+)}-r_{HZ}^{(+,+,+)}\right)^2 + \omega_3\left(\rho_{WZ}^{(+,+,+)}-r_{WZ}^{(+,+,+)}\right)^2$$
$$(3.78)$$

where $r_{HW}^{(+,+,+)}$, $r_{HZ}^{(+,+,+)}$, $r_{WZ}^{(+,+,+)}$ are the sample linear correlation coefficients (see Appendix B for further details), and the sum of the weights ω_1, ω_2 and ω_3 is unitary. In Eq. (3.78), the only dependent parameter is θ_{hwz}, because all the other parameters have already been evaluated using equations similar to Eqs. (3.71)-(3.76).

For the other cases of bivariate density functions, evaluation can be carried out by setting the sample linear correlation coefficient equal to Eq. (3.78).

3.2.4. The Meta-Gaussian Model

Multivariate probability distributions (similar to that proposed in Section 3.2.3) have the advantage of explicit formulas for the moments. For example, Eq. (3.77) represents the relationship among the linear correlation coefficient and the parameters.

Parameters $\left(\eta_h, \eta_z\right)$ are also involved in this relationship. Consequently, it is very difficult to understand the influence of the marginal distributions on the joint distribution and on the evaluation of the parameter θ_{hwz} (Salvadori et al. 2007). Moreover, analytical expressions can be obtained only for a specific set of marginal distributions. Thus, in a multivariate analysis, it is very difficult to consider different kinds of marginal functions when the fitting of all the random variables is not good if only a particular class of univariate distributions is used.

To overcome these shortcomings, an alternative methodology, the Meta-Gaussian model (Kelly and Krzysztofowicz 1997; Herr and Krzysztofowicz 2005), can be used.

In the following, without loss of generality, only the bivariate case is described. The method can be extended to the multivariate case.

Firstly, let Q and q represent the standard normal cumulative distribution and the standard normal density function, respectively (see Appendix A for further details). Moreover, let $F_H^{(+,+)}(h)$ and $F_Z^{(+,+)}(z)$ be the marginal conditional distributions related to the random variables H and Z (for simplicity, the notation for superscripts and subscripts are removed). Further, the normal quantile transform (NQT) is defined for each conditional random variable $\left(H \mid H > 0 \cap Z > 0\right)$ and $\left(Z \mid H > 0 \cap Z > 0\right)$:

$$X = Q^{-1}\left(F_H^{(+,+)}(.)\right) \tag{3.79}$$

$$Y = Q^{-1}\left(F_Z^{(+,+)}(.)\right) \tag{3.80}$$

in which Q^{-1} is the inverse of Q. From these transformations, X and Y are standard normal random variables.

Furthermore, let B be the bivariate standard normal cumulative distribution (i.e. $B(x, y, \rho) = P\left[X \le x \cap Y \le y \mid \rho\right]$), where ρ is the theoretical linear correlation coefficient between X and Y. If $F_{H,Z}^{(+,+)}(h,z)$ is Meta-Gaussian, then it can be expressed as:

$$F_{H,Z}^{(+,+)}(h,z) = B\left(Q^{-1}\left(F_H^{(+,+)}(h)\right), Q^{-1}\left(F_Z^{(+,+)}(z)\right), \rho\right)$$ (3.81)

and its density function assumes the expression:

$$f_{H,Z}^{(+,+)}(h,z) = \frac{f_H^{(+,+)}(h) \cdot f_Z^{(+,+)}(z)}{\sqrt{1-\rho^2} \cdot q\left(Q^{-1}\left(F_Z^{(+,+)}(z)\right)\right)} \cdot q\left(\frac{Q^{-1}\left(F_Z^{(+,+)}(z)\right) - \rho \cdot Q^{-1}\left(F_H^{(+,+)}(h)\right)}{\sqrt{1-\rho^2}}\right)$$

(3.82)

The conditional distribution of Eq. (3.22) requires the knowledge of the function $F_{H|Z}^{(+,+)}(h\,|\,z)$, which can be written as (Herr and Krzysztofowicz 2005):

$$F_{H|Z}^{(+,+)}(h\,|\,z) = Q\left(\frac{Q^{-1}\left(F_H^{(+,+)}(h)\right) - \rho \cdot Q^{-1}\left(F_Z^{(+,+)}(z)\right)}{\sqrt{1-\rho^2}}\right)$$ (3.83)

The coefficient ρ can be estimated from the joint sample data $\{(h_i, z_i), i = 1, ..., N_{11}\}$ of variables $(H\,|\,H > 0 \cap Z > 0)$ and $(Z\,|\,H > 0 \cap Z > 0)$. Firstly, empirical estimations $\hat{F}_H^{(+,+)}(h)$ and $\hat{F}_Z^{(+,+)}(z)$ of $F_H^{(+,+)}(h)$ and $F_Z^{(+,+)}(z)$ can be carried out by considering the subsamples $\{h_i\}$ and $\{z_i\}$ and by using the plotting position method (see Appendix B for further details). Then, for each couple (h_i, z_i) the corresponding couple (x_i, y_i) is estimated, with $x_i = Q^{-1}\left(\hat{F}_H^{(+,+)}(h_i)\right)$ and $y_i = Q^{-1}\left(\hat{F}_Z^{(+,+)}(z_i)\right)$. The resultant joint sample of observations $\{(x_i, y_i), i = 1, ..., N_{11}\}$ is then used to evaluate ρ (see Appendix B for further details).

The meta-Gaussian distribution can be validated by testing if the distribution of (X,Y) is bivariate standard normal. For this purpose, three conditions must be satisfied (see Herr and Krzysztofowicz 2005, and Appendix B for further details):

(1) Linearity of the regression between the variables Y on X;

(2) Homoscedasticity of the residual $\Theta = Y - \rho \cdot X$;

(3) The distribution of Θ has to be normal with mean equal to 0 and variance equal to $1 - \rho^2$.

To sum up, as reported in Herr and Krzysztofowicz (2005), a Meta-Gaussian model is characterized by the following features:

(1) the marginal functions $F_H^{(+,+)}(h)$ and $F_Z^{(+,+)}(z)$ can assume any mathematical expression, and they can be evaluated separately from the parameter ρ;

(2) distributions $F_{H,Z}^{(+,+)}(h,z)$, $f_{H,Z}^{(+,+)}(h,z)$ and $F_{H|Z}^{(+,+)}(h \mid z)$ are expressed in analytic form;

(3) Q, Q^{-1} and B can be evaluated numerically using the numerical tools in the technical literature (Abramowitz and Stegun 1970; Owen 1956);

(4) an easy estimation of the dependence parameter ρ (see Appendix B for further details).

3.2.5. Copula Functions

A Meta-Gaussian model, described in Section 3.2.4, is not suitable in many cases, and the use of other types of multivariate distributions may also present the following problems:

(1) some of these can only be applied to a limited range of correlation coefficient;

(2) all models are characterized by the marginal distributions belonging to the same class, but variables may show a best fit by adopting other marginal distributions.

Several works in hydrologic topics (Poulin et al. 2007; Serinaldi 2008; Villarini et al. 2008; Kuhn et al. 2007) suggest alternative structures of dependence. They are all related to the theory of Copula functions (Joe 1997;

Nelsen 2006; Salvadori et al. 2007; Serinaldi 2009), including in particular the Meta-Gaussian model (also known as the Gaussian copula).

In general, Copula functions constitute an efficient tool for multivariate analysis. Their efficiency is related to the possibility of studying marginal distributions separately (i.e. each distribution can be of a different form) and global dependence.

A brief overview (see Joe, 1997; Nelsen, 2006; Salvadori et al., 2007 for further details) is reported herein. The theory is first described for the bivariate case (2-copula), and then extended to the d-dimensions.

A 2-copula is a function $C : [0,1] \times [0,1] \to [0,1]$, characterized by the properties:

(1) $\forall s_1, s_2 \in [0,1]$, $C(s_1, 0) = C(0, s_2) = 0$, $C(s_1, 1) = s_1$ and $C(1, s_2) = s_2$

(2) $\forall s_1', s_1'', s_2', s_2'' \in [0,1]$ such that $s_1' \leq s_1''$ and $s_2' \leq s_2''$:

$$C(s_1'', s_2'') - C(s_1', s_2'') - C(s_1'', s_2') + C(s_1', s_2') \geq 0 \qquad (3.84)$$

As mathematical background, Sklar's theorem (Sklar 1959) must be mentioned: "let $F_{X_1,X_2}(x_1, x_2)$ be a joint CDF with marginal $F_{X_1}(x_1)$ and $F_{X_2}(x_2)$, then there is a copula C such that, $\forall x_1, x_2 \in R$:

$$F_{X_1,X_2}(x_1, x_2) = C[F_{X_1}(x_1), F_{X_2}(x_2)] \qquad (3.85)$$

If $F_{X_1}(x_1)$ and $F_{X_2}(x_2)$ are continuous then C is unique; otherwise C is uniquely determined on $ran F_{X_1}(x_1) \times ran F_{X_2}(x_2)$ (where ran is the range of the marginal distributions). Conversely, if C is a copula and $F_{X_1}(x_1)$, $F_{X_2}(x_2)$ are CDFs, then the function $F_{X_1,X_2}(x_1, x_2)$ defined in Eq. (3.85) is a joint CDF with marginal distributions $F_{X_1}(x_1)$, $F_{X_2}(x_2)$."

Through this theorem, a bivariate distribution can be obtained by combining two marginal functions and a suitable copula.

For hydrological applications, a class of copulas mainly used is the Archimedean 2-copula, defined as:

$$C(s_1, s_2) = \varphi^{(-1)}[\varphi(s_1) + \varphi(s_2)] \qquad (3.86)$$

where $\varphi(.)$ is the *generator*, a continuous function strictly decreasing and convex from $I = [0,1]$ to $[0, \varphi(0)]$. Among the 2-copula families, which are characterized by only one parameter, the Gumbel-Hougaard and Frank families (Nelsen 2006) can be referred to for hydrological applications:

Gumbel-Hougaard Copula

$$C(s_1, s_2) = e^{-\left[(-\ln s_1)^{\vartheta} + (-\ln u_2)^{\vartheta}\right]^{1/\vartheta}} \tag{3.87}$$

with

$$\varphi(t) = (-\ln t)^{\vartheta} \tag{3.88}$$

Frank Copula

$$C(s_1, s_2) = \frac{1}{\ln \vartheta} \cdot \ln\left[1 + \frac{\left(\vartheta^{s_1} - 1\right) \cdot \left(\vartheta^{s_2} - 1\right)}{\vartheta - 1}\right] \tag{3.89}$$

with

$$\varphi(t) = -\ln\left[\frac{\left(\vartheta^{t} - 1\right)}{\vartheta - 1}\right] \tag{3.90}$$

For these functions, the parameter ϑ is an indicator of the dependence among the variables. In particular, in Eqs. (3.87-3.88), for $\vartheta \geq 1$, s_1, s_2 are independent for $\vartheta = 1$ and positively correlated for $\vartheta > 1$. In Eqs. (3.89-3.90), for $\vartheta > 0$, s_1, s_2 are independent for $\vartheta = 1$, positively correlated for $0 < \vartheta < 1$, and negatively correlated for $\vartheta > 1$.

In summary, a multivariate analysis can be carried out by defining marginal distributions and then a copula function. Appendix A contains further details on defining marginal distributions.

To define the copula function, a semi-parametric approach, named the Canonical Maximum Likelihood (CML) method (Genest et al. 1995), can be used to evaluate the parameter ϑ for the Archimedean copula. In detail, if X_1 and X_2 are two dependent random variables, then we consider the empirical CDFs $\hat{F}_{X_1}(x_{1,i})$ and $\hat{F}_{X_2}(x_{2,i})$ of the sample $(x_{1,i}, x_{2,i})$, for $i = 1, ..., n$, by using a plotting position technique (see Appendix B for further details). Starting from the PDF which is defined as:

$$f_{X_1,X_2}(x_1,x_2) = f_{X_1}(x_1) \cdot f_{X_2}(x_2) \cdot C_{12}\left[F_{X_1}(x_1), F_{X_2}(x_2)\right] \qquad (3.91)$$

in which $C_{12}(s_1,s_2) = \dfrac{\partial^2}{\partial s_1 \partial s_2} C(s_1,s_2)$, $f_{X_1}(x_1), f_{X_2}(x_2)$ are the marginal PDFs and $s_1 = F_{X_1}(x_1)$, $s_2 = F_{X_2}(x_2)$ are the marginal CDFs, the parameter ϑ is evaluated by maximizing a likelihood function:

$$L(\vartheta, s_1, s_2) = \prod_{i=1}^{n} C_{12}(s_{1,i}, s_{2,i}) \qquad (3.92)$$

with $s_{1,i} = \hat{F}_{X_1}(x_{1,i})$ and $s_{2,i} = \hat{F}_{X_2}(x_{2,i})$.

Once evaluated ϑ for each hypothesized Archimedean family, the copula function, which reproduces data with the best fit, is chosen.

For this purpose, the function K_C is considered, corresponding to the CDF of $C(s_1,s_2)$ (i.e. $K_C(t) = P[C(s_1,s_2) \leq t]$) which is defined as (Nelsen 2006):

$$K_C(t) = t - \frac{\varphi(t)}{\varphi'(t)} \qquad (3.93)$$

This function is characterized by values that are uniformly distributed. Then, the fitting of the copula function to the sample data can be shown on a QQ ("Q" stands for quantile) plot, i.e. a plot of $K_{C[F_{X_1}(x_1), F_{X_2}(x_2)]}$ versus standard uniform quantiles as shown in Figure 3.9. In detail, in the horizontal axis the values of $K_C(t)$ are represented, with $t = C\left[\hat{F}_{X_1}(x_1), \hat{F}_{X_2}(x_2)\right]$, and the vertical axis is related to the plotting positions of $K_C(t)$.

To evaluate the best fit, comparison with the bisector (the 1:1 line) is necessary.

The generalization of the copula theory in d-dimensions, with $d \geq 3$, is reported herein (Nelsen 2006). A d-copula is a function $C:[0,1]^d \to [0,1]$, characterized by the properties:

(1) $\forall \underline{s} \in [0,1]^d$, $C(\underline{s}) = 0$ if at least one coordinate of $\underline{s}$ equals zero, and $C(\underline{s}) = s_k$ if all the coordinates of $\underline{s}$ are 1 except s_k ;

(2) $\forall \underline{a}, \underline{b} \in [0,1]^d$ such that $\underline{a} \leq \underline{b}$, $V_C[(\underline{a}, \underline{b})] \geq 0$, where $V_C[.]$ is the C-volume (Nelsen 2006).

For a d-copula, the Sklar's theorem assumes the following form (Grimaldi and Serinaldi 2006): "Let $F_{X_1,\ldots,X_d}(x_1,\ldots,x_d)$ be a d-dimensional CDF with marginal distributions $F_{X_1}(x_1),\ldots,F_{X_d}(x_d)$. Then there exists a d-copula C such that for every $\underline{x} \in R^d$:

$$F_{X_1,\ldots,X_d} = C\big[F_{X_1}(x_1),\ldots,F_{X_d}(x_d)\big] \tag{3.94}$$

If $F_{X_1}(x_1),\ldots,F_{X_d}(x_d)$ are continuous, then C is unique; otherwise C is uniquely determined in $ran F_{X_1}(x_1) \times \ldots \times ran F_{X_d}(x_d)$. Conversely, if C is a d-copula and $F_{X_1}(x_1),\ldots,F_{X_d}(x_d)$ are CDFs, then $F_{X_1,\ldots,X_d}(x_1,\ldots,x_d)$ is a d-dimensional CDF with marginal distributions $F_{X_1}(x_1),\ldots,F_{X_d}(x_d)$."

The Archimedean copulas can be defined as:

$$C(\underline{s}) = \varphi^{(-1)}\left(\sum_{i=1}^{d} \varphi(s_i)\right) \tag{3.95}$$

where $\varphi(.)$ is the generator. Kimberling's theorem (Kimberling 1974) implies that d-copulas, with $d \geq 3$, is lower bounded by the so called *product copula*:

$$C(\underline{s}) = \prod_{i=1}^{n} s_i \tag{3.96}$$

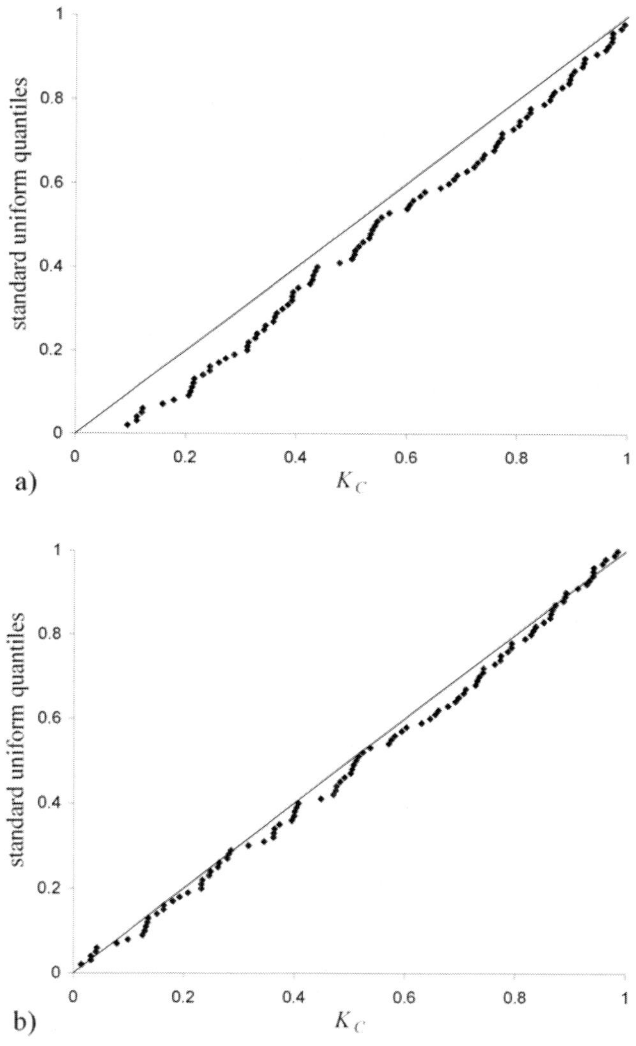

Figure 3.9. Qualitative examples of QQ plot - K_C vs standard uniform quantiles: (a) bad fit of copula; (b) good fit of copula.

This copula is related to random variables, which are independent among them. Consequently, a d-copula, with $d \geq 3$, can only reproduce positive dependence and cannot be used for negative dependence.

Extensions of Gumbel-Hougaard and Frank copulas into d-dimensions are:

Gumbel-Hougaard Copula

$$C(s_1,...,s_d) = e^{-\left[(-\ln s_1)^{\vartheta} + ... + (-\ln s_d)^{\vartheta}\right]^{1/\vartheta}}$$

(3.97)

Frank Copula

$$C(s_1,...,s_d) = \frac{1}{\ln \vartheta} \cdot \ln\left[1 + \frac{\left(\vartheta^{s_1} - 1\right) \cdots \left(\vartheta^{s_d} - 1\right)}{\vartheta - 1}\right]$$

(3.98)

In d-dimensions, the likelihood function used for CML assumes the form:

$$L(\vartheta, \underline{s}) = \prod_{i=1}^{n} C_{1...d}(s_{1,i},...,s_{d,i})$$

(3.99)

with $C_{1...d}(s_{1,i},...,s_{d,i}) = \dfrac{\partial^d}{\partial s_1 ... \partial s_d} C(s_{1,i},...,s_{d,i})$

As reported in Grimaldi and Serinaldi (2006), the d-copula will be chosen by analyzing the CDF K_C, which assumes the following form in d-dimensions (Barbe et al. 1996):

$$K_C(t) = P[C(\underline{u}) \le t] = t + \sum_{i=1}^{d-1}(-1)^i \cdot \frac{\varphi^i(t)}{i!} \cdot \kappa_{i-1}(t)$$

(3.100)

with $\kappa_i(t) = \dfrac{\kappa_{i-1}'(t)}{\varphi'(t)}$ and $\kappa_0(t) = \dfrac{1}{\varphi'(t)}$.

In order to use Copula functions for the conditional bivariate and trivariate distributions reported in Sections 3.2.1-3.2.2, it is possible to demonstrate that (De Michele et al. 2007):

$$F_{X_1|X_2}(x_1 \mid x_2) = P[X_1 \le x_1 \mid X_2 = x_2] = \frac{\partial}{\partial s_2} C(s_1, s_2) = \partial_2 C(s_1, s_2)$$

(3.101)

$$F_{X_1|X_2,X_3}(x_1 \mid x_2, x_3) = P[X_1 \le x_1 \mid X_2 = x_2 \cap X_3 = x_3] = \frac{\partial_{2,3}C(s_1, s_2, s_3)}{\partial_{2,3}C(1, s_2, s_3)}$$

$$(3.102)$$

3.3. ARTIFICIAL NEURAL NETWORKS

As explained in the beginning of this chapter, Artificial Neural Networks (ANNs) belong to the data-driven class of models. A model of this type is only dependent on the available data to be "learned". There is no "a priori" hypothesis regarding the relationships among the variables.

This type of model is a powerful tool to solve complex problems. An ANN organizes all elaborations like the human nervous system by distributing the analyses to small units, namely artificial neurons, or nodes, which are connected together. Nodes with similar characteristics are organized into one layer, and they have connections to the nodes in the other layers, but there is no interconnection. Figure 3.10 shows a simple ANN, which consists of three layers of nodes (Luk et al. 2001).

In general, there are three types of layers:

(1) the input layer - which comprises the input variables;
(2) the output layer - which comprises the output variables;
(3) the hidden layers - the layers between the input and output layers, and there can be more than one hidden layer.

Connections among the nodes of the layers allow the transfer of information from one layer to another layer. Examples of applications of ANNs for rainfall prediction are reported in French et al. (1992), Tohma and Igata (1994), and Govindaraju (2000a, b).

In the Tohma and Igata's application, three-layer ANNs based on visible and infrared remote sensing cloud images are used to predict rainfall fields. Other studies include Navone and Ceccato (1994) for forecasting summer monsoon rainfall in India, Hsu et al. (1996, 1997) for transforming satellite infrared images to rainfall rates over a catchment, Kuligowski and Barros (1998), who used upper atmospheric wind direction and antecedent rainfall as input data to generate forecast precipitation for the successive 6 hours, Lin and

Chen (2005), Luk et al. (2000, 2001) and Ramirez et al. (2005) for rainfall predictions in Taiwan, Australia and Sao Paulo State in Brazil, respectively.

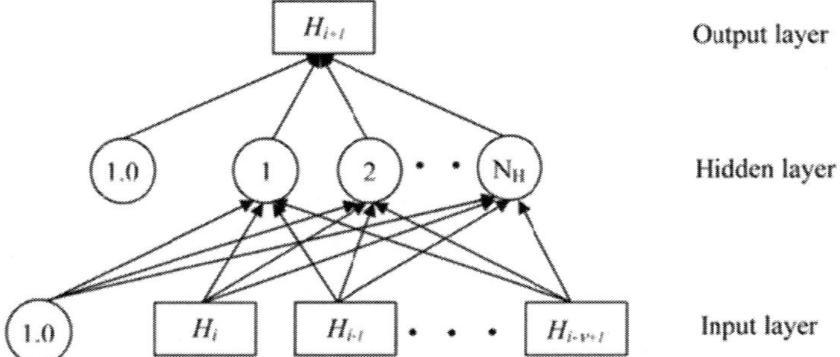

Figure 3.10. Example of an ANN (adapted from Luk et al. 2001).

In general, an ANN designed for rainfall nowcasting, may be schematized by the following Markovian expression:

$$\underline{H}_{i+1} = g\left(\underline{H}_i, \underline{H}_{i-1}, \underline{H}_{i-2}, ..., \underline{H}_{i-v+1}\right) + e_{i+1} \qquad (3.103)$$

in which $\underline{H}_{i+1}$ is a vector of rainfall heights $H_{i+1,1}, H_{i+1,2}, ..., H_{i+1,N}$ at N different locations and cumulated over the interval $\left[i\Delta t; (i+1)\Delta t\right]$, $g(.)$ is a non-linear function, e_{i+1} is an error to be minimized and v, named lag, is the (unknown) number of antecedent rainfall to be considered. In this context, the development of an ANN comprises:

- selection of a network scheme suitable for rainfall prediction;
- evaluation of the lag v;
- assessment of the best network complexity (i.e. the number of hidden layers, the number of nodes in each hidden layer and their related parameters).

In technical literature, there are many proposed ANNs structures. A simple structure of ANN is constituted by the multilayer feedforward network (MLFN), which is described in Section 3.3.1 and in Luk et al. (2001). Section 3.3.2 concerns calibration of an ANN. In Section 3.3.3, the methodology for carrying out a PQPF with ANNs is explained.

3.3.1. Multilayer Feedforward Network (MLFN)

If the information is only transmitted from one layer to another layer in the forward direction, then the network is called a feedforward network, or a Multilayer Feedforward Network (MLFN).

Figure 3.10 shows a simple three layer MLFN for rainfall prediction. It is characterized by v input nodes (i.e. the v values of antecedent rainfall heights), N_H hidden nodes, and one output node (i.e. the predicted rainfall height in the forecasting time interval), and it takes the form:

$$H_{i+1} = S_1\left(\sum_{n=1}^{N_H} O_n \cdot w_n + w_0\right) \qquad (3.104)$$

where:

- H_{i+1} is the output of the ANN;

- O_n is a function evaluated for the nth hidden node, defined as:

$$O_n = S_2\left(\sum_{k=1}^{v} H_{i+1-k} \cdot w_{kn} + w_{0n}\right) \qquad (3.105)$$

- H_{i+1-k} are the inputs of the MLFN, with $k = 1,...,v$;
- w_n are the connection weights between nodes of the hidden and output layers;
- w_{kn} are the connection weights between nodes of the hidden and the input layers;
- 1.0 is a bias and w_0 and w_{0n} are the weights for the biases;
- S_1 and S_2 are named *activating functions*. For example a logistic sigmoid function is usually used:

$$S(x) = \frac{1}{1 + e^x} \qquad (3.106)$$

This function may be adopted for the hidden nodes, aimed at allowing nonlinearity in the network. However, it is not suitable for the output nodes, as all the rainfall heights would assume values between 0 and 1. Thus, for the output nodes an identity linear function is usually assumed.

Extension of Eqs. (3.104-3.105) to the case of space-temporal prediction is straightforward. Figure 3.11 shows a generic structure in which:

- the output nodes are the rainfall heights during the time step $(i+1)$ for N rain gauges;
- the input layers contain v set of input nodes, and each set groups the rainfall heights of N rain gauges at time step $(i-k+1)$, with $k=1,...v$.

Of course, for both cases (temporal and spatiotemporal models) the number of hidden layers, the number of hidden nodes for each hidden layer, and the lag v, must be determined.

3.3.2. ANN Calibration

The selection of the best network complexity for a given time series is a lengthy procedure. It requires many trials and comparisons among several configurations with different numbers of hidden layers, hidden nodes, and values of lag v. For each configuration to be tested, the parameters to be estimated are the weights w_n, w_{kn}, w_0 and w_{0n}.

In general, an ANN has a good generalization if it "learns" all the data features. The complexity of an ANN structure influences this learning process. In fact, if the structure is too simple, then an ANN is not able to reconstruct all the features, and then an underfitting of data emerges. On the contrary, if the structure is too complex, then noise may be also reproduced, providing an overfitting of data.

In the calibration phase, a particular split-sample approach named early stopping method (Luk et al. 2001; Sarle 1995) can be adopted. The data are organized into three sets: (1) training, (2) monitoring and (3) validation sets. The training set is used to evaluate the parameters. The monitoring set is used to check the model performance at regular intervals during the training. By considering only the monitoring set, training is stopped when the error is minimized. The validation set is used for final evaluation of the model capability to reproduce the features of rainfall data.

In order to obtain a fast convergence in a calibration analysis, Luk et al. (2001) and Nasseri et al. (2008) proposed the following transformation of original data:

$$y = 0.5 \cdot \log_{10}(H + 1) \qquad (3.107)$$

in which H is the rainfall data, y is the data after the transformation.

To assess the goodness of fit between the observed and predicted data, the Normalized Mean Squared Error (NMSE) may be used (Luk et al. 2001):

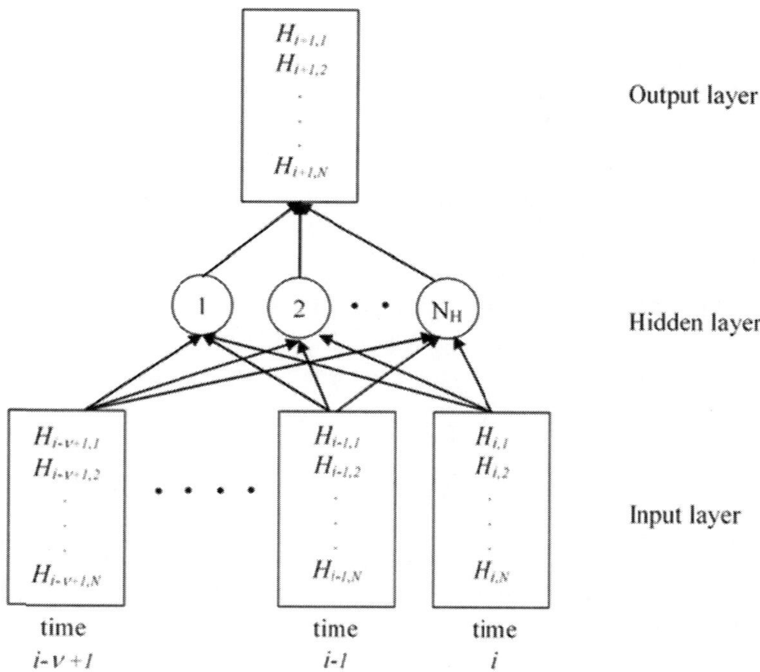

Figure 3.11. Example of a MLFN for a spatiotemporal domain (adapted from Luk et al. 2001).

$$NMSE = \frac{\displaystyle\sum_{o=1}^{O}\sum_{p=1}^{P}\left(y_{op} - y_{op}^{*}\right)^{2}}{\displaystyle\sum_{o=1}^{O}\sum_{p=1}^{P}\left(y_{op} - m_{y_{op}}\right)^{2}} \qquad (3.108)$$

where O is the total number of output nodes, P is the total number of analyzed temporal data for each output node, y_{op} is the transformed observed rainfall data, $m_{y_{op}}$ is the expected value of the transformed observed data and y^{*}_{op} is the transformed network outputs. If $NMSE = 1$, then there is only the prediction of the expected value $m_{y_{op}}$.

Many optimization algorithms can be used for Eq. (3.109). For example, the local optimization algorithm such as Backpropagation (Maier and Dandy 2000), the global search methods such as Shuffled Complex Evolution (SCE , Duan et al. 1992), and the Genetic Algorithm (GA, Montana and Davis 1988; Maniezzo 1994) can all be adopted.

To sum up, various network configurations can be attempted for a given set of data (with different number of hidden layers, nodes, and lag v) in order to determine the scheme which gives the best performance.

It must be pointed out that this procedure only gives a deterministic forecast. As mentioned in Chapter 2, a probabilistic forecast is preferred. Moreover, with the inherent variability of the rainfall process, different data sets actually yield different "optimal" parameter values. With these shortcomings, Section 3.3.3 describes a methodology that takes the parameter uncertainty into account, thereby giving a probabilistic forecast.

3.3.3. Probabilistic Forecasting with ANN

The nonlinearity of ANNs implies the presence of multiple optima solutions. Consequently, many combinations of network weights may provide similar performances.

The method described in Section 3.3.2 does not guarantee that the global solution of the network will be found. Thus, a more robust model has to be developed in order to take into account the parameter uncertainty.

For this purpose, Kingston et al. (2003) proposed a Bayesian methodology, in which a conditional probability distribution is adopted for the vector of parameters:

$$P[\underline{w} \mid D, M] = \frac{P[D \mid \underline{w}, M] \cdot P[\underline{w} \mid M]}{P[D \mid M]} \qquad (3.109)$$

where $\underline{w}$ is a vector of model parameters (i.e. the connection weights), M is the set of model predictions and D represents the data. The *likelihood function* $P[D \mid \underline{w}, M]$ is obtained by comparing the measurements with the model predictions. It is the function that updates the *prior function* $P[\underline{w}\,|M]$ of $\underline{w}$ by considering the data.

The *prior function* is the density probability distribution of the model parameters, obtained without the knowledge of data. For example, it can be assumed as a multivariate uniform distribution, in which each set of parameter values has the same probability of occurrence.

To sum up, the calibration of a Bayesian, or stochastic, ANN, considers vectors of parameter values from the *posterior distribution* $P[\underline{w}\,|\,D,M]$, rather than a single "optimal" set.

Markov Chain Monte Carlo method (MCMC, Neal 1992; Thyer et al. 2002) can be adopted to evaluate $P[\underline{w}\,|\,D,M]$ numerically for high dimensions of $\underline{w}$.

If samples of network weights are generated from $P[\underline{w}\,|\,D,M]$ and new data are the input into an ANN, a distribution of the network output values will be obtained. Consequently, confidence intervals may also be determined, and thus similar plots like Figures 3.3 and 3.8 can be obtained.

CONCLUSION

In this chapter, the background of several kinds of stochastic models, as proposed in technical literature, is described. In particular, it is noted that the model driven class comprises all the approaches suitable for Probabilistic Quantitative Precipitation Forecast (PQPF). By using ARMA or other models based on mixed distributions, it is possible to obtain the exceedence or non-exceedence probability of given threshold values, or forecasts of particular exceedence or non-exceedence percentiles. ANNs, belonging to the data driven class, are proposed in literature with a typical framework for Quantitative Precipitation Forecast (QPF). In Section 3.3.3, a Bayesian methodology is described such that ANNs can be used for probabilistic nowcasting.

For a model developer, it is very difficult to establish a priori the kind of model which is most suitable for a given dataset. To select the "best" model, one procedure is to compare the model performances during validation phase.

Chapter 4

METEOROLOGICAL MODELS

In this chapter, a brief overview of meteorological models, also named as Numerical Weather Prediction (NWP) models, is provided. The mathematical background is constituted by the fundamental equations of fluid mechanics (Section 4.1).

To solve these equations is an initial and boundary value problem (IBVP). The theoretical equations, defined onto a continuous spatial and temporal domain, are numerically solved on a discrete domain (i.e. a grid of spatial points at fixed time instants). In order to solve them, a parameterization is also necessary.

This parameterization is related to all the physical processes at a finer scale, which are not solved explicitly using the grid of calculus. Consequently, the accuracy of NWP depends on the accuracy of initial and boundary conditions, the discretization of domain and the parameterization. As mentioned in Chapter 1, even a high resolution NWP is still too coarse to be used alone for hydrological applications, which require small temporal and spatial scales. Consequently, at these small scales, it is necessary to use stochastic or coupled meteo-stochastic models as described in Chapter 5. Section 4.2 contains a classification of meteorological models, which can be regrouped into Global Circulation Models (GCMs) and Limited Area Models (LAMs). Section 4.3 focuses on the error sources and the importance of data assimilation (i.e. the updating of observed data during a model run in real time). In order to take into account the uncertainties of initial state of atmosphere, Section 4.4 explains the ensemble forecasting techniques. In Section 4.5, as an example of NWP model, the MM5 model is described. The coupling and the application of the MM5 model with a stochastic model is described in Chapter 5.

4.1. GOVERNING EQUATIONS

A NWP model comprises partial differential equations, which describe the dynamics of the atmosphere mathematically. In the following, only a qualitative description of these equations is provided without the mathematical details. This approach makes it easier for the readers to understand the concepts inherent in the modelling.

Input data are constituted by measures in meteorological stations, relating to pressure, wind velocity, temperature and air humidity for a thickness of atmosphere equal to about 20–50 km.

These quantities are indicated as *prognostic variables*, and they are the most important, as they are directly measured and allow evaluation of all the other quantities of interest, named *diagnostic variables* such as rainfall heights, cloudiness, etc.

The World Meteorological Organization (WMO) acts to standardize the instrumentation, observing practices and timing of these observations worldwide. A variety of instrumentations are used to gather observational data for use in NWP models. Radiosondes are adopted, which rise through the troposphere and well into the stratosphere. Moreover, information from weather satellites is used where traditional data sources are not available. Meteorological data are also available from aircraft and shipping routes. As an example, research projects use reconnaissance aircraft to analyze particular weather systems, such as tropical cyclones.

The most important partial differential equations for the prognostic variables, considered by all the NWP models, are:

1. *Navier Stokes equations.* These equations model the three-dimensional wind field. They were derived by applying Newton's second law to fluid motion, together with the assumption that the fluid stress is the sum of a diffusing viscous term (proportional to the gradient of velocity), plus a pressure term.
2. *Thermodynamic equation.* This equation is the fundamental first law of thermodynamics, also named as equation of energy conservation.

3. *Water Balance equation.* This equation takes into account all the processes of water balance, such as evaporation, condensation, fusion, solidification and sublimation.

4. *Continuity equation.* This equation ensures that, in a fixed control volume, the input quantity of air is the sum of the output quantity plus the accumulated one in a determinate time interval.

For the diagnostic variables, there are two more equations:

1. *Gas State equation.* This equation models the relationships among pressure, density, temperature and volume of an air mass.

2. *Hydrostatic equation.* This equation accounts for the variation in pressure with elevation and air density.

This system of partial differential equations is nonlinear. Thus, except for a few idealized cases, it is impossible to solve them exactly through analytical methods. Hence, numerical methods are used and approximate solutions are obtained. Different models use different solution methods. Some models use spectral methods for the horizontal dimensions and finite difference methods for the vertical dimension, while others use finite-difference methods in all three dimensions. In general, a discretization of spatial domain is carried out, which implies a definition of a grid of points where each variable is identified and forecasted. Each point (or node) is representative of a portion of atmosphere. For an assigned spatial domain, it is apparent that if there are more nodes then the numerical solution is more accurate, but the computational costs (in terms of time required to carry out the forecasting for the domain) are higher. This is the main aspect to be considered when a NWP model is developed, and it is intuitive that the grid resolution has to be coarser for greater spatial domains (see Section 4.2).

4.1.1. Coordinate Systems

In general, NWP models do not adopt a Cartesian system. The horizontal grid is referred to cartographic projections like Polar Stereographic, Lambert Conformal and Mercator ones, respectively for high, medium and low values of latitude. Moreover, to facilitate the computation of numerical solutions, staggering configurations are adopted. It means that different grids are used for the variable. As an example, Figure 4.1 shows the horizontal Arakawa-Lamb

B-grid staggeringused for the Fifth-Generation Mesoscale Model (MM5, see Section 4.5), in which temperature (T), pressure (p) and other scalar quantities are referred to the center of the grid square, while the eastward (u) and northward (v) velocity components are evaluated at the corners.

The vertical coordinate is usually "terrain following" (Figure 4.2). It means that the lower levels assume the form following the terrain surface, while the uppermost level is horizontal. Intermediate levels from lower levels to the uppermost level then gradually change from terrain following to horizontal with pressure decreasing. For a terrain following system, the quantity σ is defined as:

$$\sigma = \frac{p - p_{top}}{p_{s0} - p_{top}} \tag{4.1}$$

where p is pressure at the generic level, p_{top} is the pressure value at the top level, p_{s0} is the pressure value at the surface. From Eq. (4.1), it is apparent that σ equals zero at the top level, and equals one at the surface. It is not necessary to have regularly spaced levels between the surface and the top level. In general, the resolution in the lower layer is much finer than those at the upper layer.

4.1.2. Parameterization

As mentioned at the beginning of this chapter, a parameterization is necessary, relating to all the physical processes at a finer scale, which are not explicitly solved by considering the grid of calculus. In fact, a NWP model evaluates only average values into each pixel, but it has to take into account the variability inside the cell. As such, algorithms are developed, which describe phenomena at a finer scale on the basis of the average values. Parameterization is usually related to the following processes:

1. *Radiative transfer:* absorption of short wave radiation, absorption and emission of long wave radiation in the atmosphere, cloud effect.
2. *Soil-atmosphere interaction:* absorption of short wave radiation, absorption and emission of long wave radiation by the Earth's surface, sensible and latent heat flows, role of vegetation.

3. *Turbulence:* description of heat and humidity fluxes induced at scales finer than the model resolution.

4. *Convection:* dynamic and thermodynamic effects of the convective motions not explicitly resolved by the partial differential equations of a NWP model.

5. *Microphysical processes:* phase transitions of water vapor, water and ice in the atmosphere, formation and deposition of hydrometeors (rain, snow, hail).

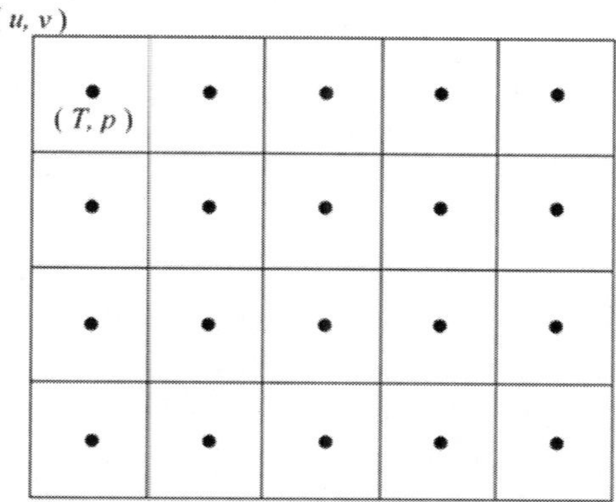

Figure 4.1. Horizontal Arakawa-Lamb B-grid staggering used for the MM5 model (adapted from http://www.mmm.ucar.edu/mm5/documents/tutorial-v3-notes-pdf/intro.pdf).

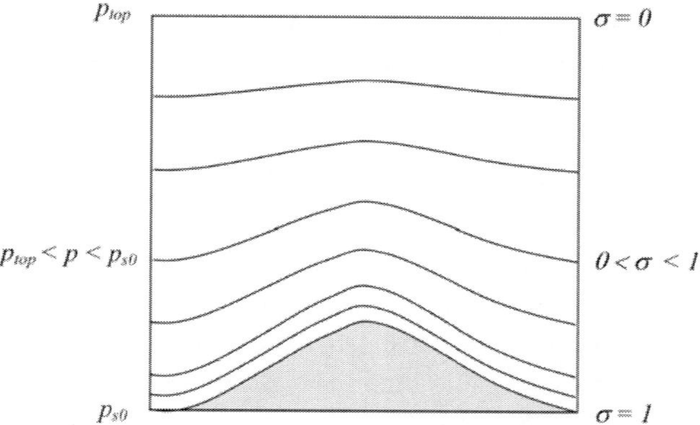

Figure 4.2. Representation of "terrain following" vertical coordinates (adapted from http://www.mmm.ucar.edu/mm5/documents/tutorial-v3-notes-pdf/intro.pdf).

4.2. CLASSIFICATION OF METEOROLOGICAL MODELS

In the technical literature and also on the Internet websites, there are many NWP models descriptions. These models can be distinguished by their applicability to different geographical conditions, such as topography and vegetation. They can also be distinguished by the different numerical methods adopted, or by the diversity in the treatment of physical processes and parameterizations, such as turbulence, convection and surface fluxes. The development of different models allows the so-called multi-model forecast, which compares output values of several schematizations (see Section 4.4 for details). In general, all the meteorological models can be grouped in two main classes:

- *Global Circulation Models (GCMs).* They consider a spatial domain covering the entire Earth. Consequently, in order to reduce computational costs, the spatial resolution has to be coarse, typically equal to 40-100 km in the horizontal direction. The temporal resolution is typically equal to 6-12 hours. The forecasting is provided up to 384 hours in advance. Due to the coarse resolutions, GCMs can only give approximate solutions.
- *Limited Area Models (LAMs).* They cover only a part of the Earth space, and consequently, finer resolutions are adopted. In details, the spatial resolution is typically equal to 1-20 km in the horizontal direction, and the temporal resolution is typically equal to 1-3 hours. The forecasting is provided up to 72 hours in advance.

It must be highlighted that GCMs are often used to initialize LAMs. In fact, before starting a run, a LAM adopts the output values from a GCM as initial conditions. Moreover, GCMs provide boundary conditions to LAMs for the entire forecasting period. It is obvious that there will be some "holes", as some of the initial and boundary conditions on the finer LAM grid are not available from the GCM outputs. As such, a suitable interpolation technique is required to determine the missing initial and boundary conditions. Tables 4.1-4.2 contain respectively a list of GCMs and a list of LAMs.

Another classification is related to the mathematical simplifications, which neglects some parameters. For NWP models, it is apparent that the horizontal dimension of atmospheric cyclones (typically 1000 km) is much larger than the vertical dimension (typically 10 km) and the motion is predominantly horizontal.

Table 4.1. List of some Global Circulation Models (GCMs)

CGM	developed by
Global Forecast System (GFS)	National Oceanic and Atmospheric Administration (NOAA)
Navy's Operational Global Atmospheric Prediction System (NOGAPS)	US Navy
Global Environmental Multiscale (GEM)	Meteorological Service of Canada (MSC)
Integrated Forecast System (IFS)	European Centre for Medium-Range Weather Forecasts (ECMWF)
Unified Model (UM)	UK Met Office
Action de Recherche Petite Echelle Grande Echelle (ARPEGE)	French Weather Service Météo-France
Intermediate General Circulation Model (IGCM)	Department of Meteorology, University of Reading, Department of Atmospheric and Oceanic Sciences at McGill University

Consequently, the vertical acceleration of air is much less than the acceleration due to gravity. Hence, the air acceleration can be neglected together with the friction forces. This assumption is known as the hydrostatic approximation. All the models that use this assumption are called hydrostatic models. They include all GCMs and a subset of LAMs, characterized by a horizontal spatial resolution up to 5 km. Hydrostatic approximation enables simpler calculations thereby reducing the computational costs. However, it prevents the modelling of violent phenomena at small scales, such as thunderstorms and tornadoes, for which the vertical motions (convection) play a crucial role. Thus, LAMs with finer resolutions (1-3 km) do not use hydrostatic approximations, and they are called not-hydrostatic models.

4.3. ERROR SOURCES

There are two main error sources for a NWP model:

1. *Model errors.* These errors are related to model structure such as limited spatial resolution, approximations about parameterization schemes, numerical errors, and physical phenomena that are not described by the model.

2. *Initial condition errors.* This is the most important error source. It is induced by observed data, which are insufficient or irregularly distributed or of poor quality, and lack of measurements for some parameters such as soil moisture.

Table 4.2. List of some Limited Area Models (LAMs)

LAM	developed by
Weather Research and Forecasting (WRF)	National Centers for Environmental Prediction (NCEP), National Center for Atmospheric Research (NCAR)
Regional Atmospheric Modeling System (RAMS)	Colorado State University
Mesoscale Model of 5th generation (MM5)	Pennsylvania State University (PSU) National Center of Atmospheric Research (NCAR)
Advanced Region Prediction System (ARPS)	University of Oklahoma
High Resolution Limited Area Model (HIRLAM)	Cooperation among several European meteorological institutes
Global Environmental Multiscale Limited Area Model (GEM-LAM)	Meteorological Service of Canada (MSC)
High Resolution Numerical Weather Prediction Project (ALADIN)	Several European and North African countries under the leadership of Météo-France
Consortium for Small-scale Modeling (COSMO) formerly known as LAMI	German Weather Service Climate Limited-area Modelling (CLM)-Community
MESOscale Non-Hydrostatic model (MESO-NH)	Centre National de Recherches meteorologiques (CNRM Meteo France) Laboratoire d'aerelogie (LA)
Bologna Limited Area Model (BOLAM)	Institute of Atmospheric Sciences and Climate (ISAC) of the Italian National Research Council (CNR)
Mpdello Locale in H coordinate (MOLOCH)	Institute of Atmospheric Sciences and Climate (ISAC) of the Italian National Research Council (CNR)

As dynamics of atmosphere is represented by a non-linear system of equations, errors in initial conditions tend to amplify with increasing forecast horizon. Consequently, the data assimilation process, as described in Section 4.3.1, assumes a crucial role to improve the quality of forecasting.

4.3.1. Data Assimilation

The first step to forecast the future state of atmosphere is the determination of the present state, which is based on the available observed data. These data are generally provided by an irregularly spatial distributed network of sensors. However, a homogeneous knowledge of the initial state is required. In the case of GCMs, it means the entire Earth. It is therefore necessary to extend the irregular spatial information to all the nodes of the discretized domain for a given study area.

With this aim, it is possible to adopt mathematical interpolation techniques (Panofsky 1949; Gilchrist and Cressman 1954), or more complex methods that model physical and statistical properties of the atmospheric field.

With the more complex methods, for a fixed time instant before the forecasting, the basic idea is to carry out "a priori" estimations of the quantities of interest, such as temperature, pressure, wind velocity, etc. for the study area. These estimations are derived from a previous run of the model, and then adjusted to match the observed data. The "a priori" estimation of a field is called Back-Ground Field (BGF). The main advantage of this method, known as data assimilation, is to initialize a NWP model by taking into account both measurements at the initial time and previous observed data.

Techniques for data assimilation are constituted by Successive Corrections (Bergthorsson and Doors 1955; Creessman 1959) and Optimal Interpolation (Gandin 1963). Both techniques calculate, in an assigned node, the increment to add algebraically to the BGF. In brief, for each point where there is a sensor, the differences (named as innovations) between observed data and BGF are evaluated. Then the increment is equal to a weighted summation of all these innovations.

4.4. ENSEMBLE PREDICTION SYSTEMS

As mentioned in Section 4.3, initial conditions are the most important source of errors in NWP models. Lorenz (1963) demonstrated that atmospheric systems are affected by a significant dispersion of the results if initial conditions are truncated to seven decimal places. In other words, the non-linear structure of equation system for atmospheric dynamics is very sensitive to small variations in initial conditions. This includes the initial conditions derived using data assimilation as described in Section 4.3.1. For this reason, atmosphere is known as a chaotic system.

Due to this problem, in recent years, techniques of ensemble prediction have been developed. With this technique, using an assigned meteorological model, several runs with slightly different initial conditions are performed. Consequently, several future scenarios are obtained, which constitute an ensemble.

As many model simulations must be carried out for the same forecasting period, the computational cost therefore plays a crucial role. As such, it is important to choose initial conditions that are admissible initial states of atmosphere based on the following two fundamental requirements:

1. The sample of initial conditions has to be representative of the realistic probability distribution of errors for data assimilation.
2. The resultant future scenarios, known as realizations, have to be good approximations of real atmospheric scenarios.

These requirements represent very difficult problems. First, the probability distribution of initial errors is unknown. Thus, generation of purely random initial states does not provide a realistic distribution of future states. To overcome these problems, each meteorological operative center has adopted several techniques (Singular Vector, Ensemble Kalman Filter, Breeding Vector) to generate small perturbations, thus creating the ensemble of initial states to be used to estimate future atmospheric scenarios.

Figure 4.3 shows a qualitative schematization of an ensemble prediction, by considering an ideal state of two-dimensional space. At $t = 0$, inside the first rectangle, there are both the unperturbed initial state (black circle) and the perturbed initial states (empty circles). By running a NWP model, each initial state generates a future scenario over time, and the evolution is represented by the associated trajectory. The whole set of scenarios, represented by a spaghetti plot, constitutes an approximation of the probability distribution for the evolution of the atmosphere over time.

In this qualitative example, it can be observed that at $t = t_1$, the generated states differ slightly from each other. It means that a NWP model produces similar predictions at the first step of forecast. Consequently, the uncertainty for the states at $t = t_1$ is therefore not so different to that at $t = 0$.

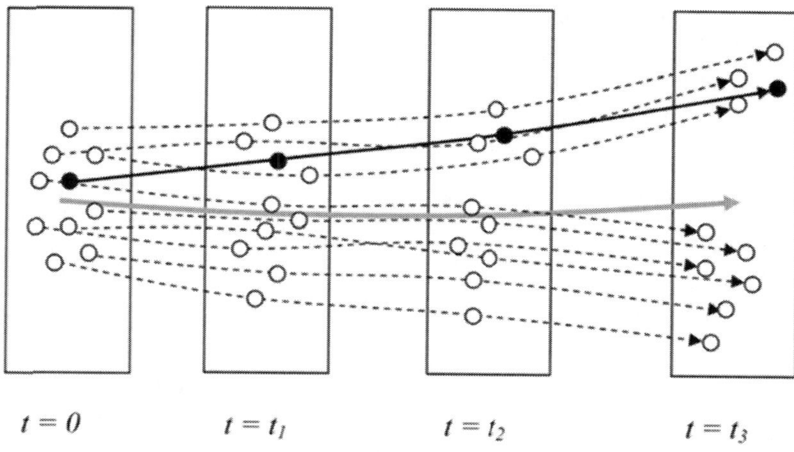

Figure 4.3. Qualitative schematization of an ensemble prediction.

However, at $t = t_2$, the realizations (trajectories) diverge from each other. Four scenarios tend towards one atmospheric state, while the other six tend towards another state. As such, the more probable future state could be considered as the one with the greater number of realizations. The spread at $t = t_2$ allows an evaluation of the forecasting uncertainty. In fact, if only the initial state without added errors is considered (black circle), the resulting forecast scenario could be the less probable state.

In general, at each time, the ensemble forecast is mainly evaluated in terms of an average scenario, known as the ensemble mean. This is to obtain a more accurate prediction of future state. This ensemble mean is different from the scenario obtained by evaluating the average value of the initial conditions and then running the NWP model. It is because of the non-linearity of the mathematical equations.

The benefit of the ensemble mean is to highlight the more probable scenario by reducing the effects of the "extreme" future states. In general, an ensemble mean is reliable unless a high divergence occurs for the predicted future states. This is the case at $t = t_3$, in which the obtained mean (grey line in Figure 4.3) is not representative of any cluster of states.

Ensemble prediction systems can be carried out by considering the output of several NWP models. In this case, a multi-model forecast, or multi-model ensemble, is obtained.

It must be pointed that in the technical literature, the ensemble prediction systems are also suggested for probabilistic forecast in terms of PQPF (Seo et al. 2000; Krzysztofowicz 2001). However, they underestimate the total uncertainty because not all sources of error are considered (Antolik 2000; Golding 2000).

4.5. MM5 MODEL

This section contains a brief overview of the MM5 model. Its output values in terms of rainfall heights were used by the author of this book for coupling with a stochastic model (see Section 5.2).

The MM5 model is the Fifth-Generation NCAR/Penn State Mesoscale Model, and it is the latest of a series developed by Pennsylvania State University (PSU) and by National Center of Atmospheric Research (NCAR) (Anthes and Warner 1978). It is a non-hydrostatic LAM, includes multiple-nest and data assimilation capabilities. Further details of the model can be found in http://www.mmm.ucar.edu/mm5/documents/. The following is a summary of the main properties:

- Horizontal grid is an Arakawa-Lamb B-Staggering (see Section 4.1.1), while the vertical coordinate is terrain-following (see Section 4.1.1).
- It allows multiple nesting with up to nine domains. All domains can be run simultaneously. The nesting ratio is 3:1 for two consecutive levels (i.e. the coarser level in terms of space and time resolutions is three times less than the finer level). Figure 4.4 shows a qualitative example, in which it can be observed that multiple nests can be adopted at the same level (in this case domains 2 and 3). They can also be overlapped. Each domain has a mother domain. Domain 1 is the mother domain for domains 2 and 3, while domain 2 is the mother domain for domain 4.
- Meteorological data are horizontally interpolated from a latitude-longitude system to a rectangular domain (Mercator, Lambert or Polar Stereographic projection) with the modules named TERRAIN and REGRID. Module LITTLE_R/RAWINS enhances the interpolated data with observations from a standard network of surface and radiosonde stations.
- Vertical interpolation from pressure levels to σ terrain following is carried out by the module INTERPF. After a MM5 model run, the

module INTERPB is used to interpolate from σ terrain following to pressure levels.

- The module NESTDOWN is used to interpolate output values of an assigned level into a finer level.

- The modules RIP and GRAPH give the output visualization in terms of pressure and σ levels

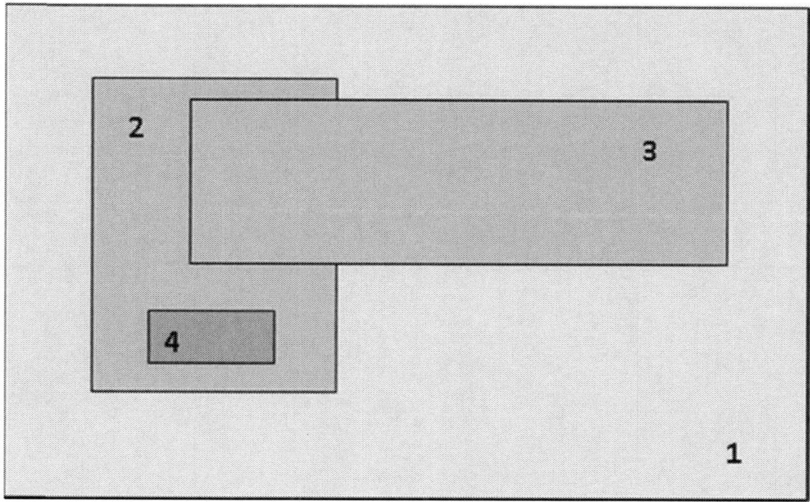

Figure 4.4. Qualitative representation of nesting configuration for MM5 model (adapted from http://www.mmm.ucar.edu/mm5/documents/tutorial-v3-notes-pdf/intro.pdf).

CONCLUSION

This chapter focused on the general aspects of the NWP models, including the governing equations, data assimilation and techniques of ensemble prediction.

Chapter 5

COUPLED METEOROLOGICAL-STOCHASTIC MODELS

As mentioned in Chapter 1, coupling stochastic models with outputs from meteorological models (Chapter 4) is useful for rainfall forecasting. The goal is to improve the forecast on the hydrological scale, especially for the case "nowcasting outside a storm event". For the deterministic meteorological forecast (or for an ensemble mean forecast), it is possible to couple them with stochastic models by using a Bayesian Forecasting System or BFS (Krzysztofowicz 1999). Similar approaches have been used to improve radar estimation of rainfall using rain gauge information (Seo and Smith 1991; Todini 2001).

In this chapter, we first desccribe the theoretical background of a BFS in Section 5.1. Then, an example of the coupled model developed at the University of Calabria, Italy is described in Section 5.2.

5.1. STRUCTURE OF A BAYESIAN FORECASTING SYSTEM (BFS)

Each hydrologic prediction, carried out by a deterministic model, is affected by uncertainty. In detail, it can be decomposed into two sources (Krzysztofowicz 1999):

1. Input uncertainty associated with random inputs to the model.

2. Hydrologic uncertainty related to all sources beyond those classified as random inputs.

The inference scheme is depicted in Figure 5.1, in which it is possible to define the following random variables:

- $\underline{U}$ is the vector comprising all the input variables which are affected by uncertainty, and for which a probability density function $\eta_{\underline{U}}(\underline{u})$ can be defined. All the deterministic input variables are not considered in this vector. In the case of a meteorological model, $\underline{U}$ is dependent on pressure, temperature, and wind velocity of each individual site to be investigated;

- $\underline{H}_{i+1}$ (predictand) is the vector column containing the rainfall height values of all the sites under investigation, which will occur over the future time interval $[i\Delta t;(i+1)\Delta t]$;

- $\underline{S}_{i+1}$ is the vector column which represents the output of the meteorological model. It is the output for each spatial pixel containing rain gauge or site under investigation and constitutes an evaluation of $\underline{H}_{i+1}$;

Moreover, a BFS is characterized by the following processors:

1. the Input Uncertainty Processor, which assumes the meteorological model as perfect. It yields the *output density* $\pi_{\underline{S}_{i+1}}(\underline{s}_{i+1})$, which describes the uncertainty associated to the input $\eta_{\underline{U}}(\underline{u})$;

2. the Hydrologic Uncertainty Processor, which evaluates an aggregate of the uncertainties arising from other sources (the model structure, parameter estimation, measurement errors of physical quantities, etc.), under the hypothesis that there is no input uncertainty. It yields the *posterior density* $\phi_{\underline{H}_{i+1}|\underline{S}_{i+1},\underline{H}_0}(\underline{h}_{i+1}|\underline{s}_{i+1},\underline{h}_0)$ of the predictand $\underline{H}_{i+1}$, conditional on the rainfall heights vector $\underline{H}_0$, observed before the forecast, and on the model output $\underline{S}_{i+1}$, estimated considering $\underline{U}$ as perfect;

3. the Integrator Processor, which takes into account both sources of uncertainty, and yields the predictive density:

$$\psi_{\underline{H}_{i+1}|\underline{H}_0}\left(\underline{h}_{i+1}\mid\underline{h}_0\right)=\int_0^{+\infty}\int_0^{+\infty}\cdots\int_0^{+\infty}\phi_{\underline{H}_{i+1}|\underline{S}_{i+1},\underline{H}_0}\left(\underline{h}_{i+1}\mid\underline{s}_{i+1},\underline{h}_0\right)\cdot\pi_{\underline{S}_{i+1}}\left(\underline{s}_{i+1}\right)\cdot d\underline{s}_{i+1}$$

$$(5.1)$$

representing a PQPF for H_{i+1}. As $\underline{S}_{i+1}$ is a vector, the predictive density has to be evaluated by resolving a multiple integral.

For simplicity, in this chapter, no input uncertainty is supposed, so the density $\psi_{\underline{H}_{i+1}|\underline{H}_0}\left(\underline{h}_{i+1}\mid\underline{h}_0\right)$ equals to the posterior density $\phi_{\underline{H}_{i+1}|\underline{S}_{i+1},\underline{H}_0}\left(\underline{h}_{i+1}\mid\underline{s}_{i+1},\underline{h}_0\right)$ (Krzysztofowicz 1999). This will be applied to subsequent sections of this chapter.

5.1.1. Hydrologic Uncertainty Processor

The posterior density $\phi_{\underline{H}_{i+1}|\underline{S}_{i+1},\underline{H}_0}\left(\underline{h}_{i+1}\mid\underline{s}_{i+1},\underline{h}_0\right)$ derives from a revision of a *prior density* $g_{\underline{H}_{i+1}|\underline{H}_0}\left(\underline{h}_{i+1}\mid\underline{h}_0\right)$, which exists before the preparation of a forecast. The revision is based on a likelihood function $f_{\underline{S}_{i+1}|\underline{H}_{i+1},\underline{H}_0}\left(\underline{s}_{i+1}\mid\underline{h}_{i+1},\underline{h}_0\right)$, calibrated from past evidence of model performance $\underline{S}_{i+1}=\underline{s}_{i+1}$ against observations $\underline{H}_{i+1}=\underline{h}_{i+1}$ and $\underline{H}_0=\underline{h}_0$. According to the Bayes' theorem (see Appendix A for further details), the general formulation of $\phi_{\underline{H}_{i+1}|\underline{S}_{i+1},\underline{H}_0}\left(\underline{h}_{i+1}\mid\underline{s}_{i+1},\underline{h}_0\right)$ takes the form:

$$\phi_{\underline{H}_{i+1}|\underline{S}_{i+1},\underline{H}_0}\left(\underline{h}_{i+1}\mid\underline{s}_{i+1},\underline{h}_0\right)=\frac{f_{\underline{S}_{i+1}|\underline{H}_{i+1},\underline{H}_0}\left(\underline{s}_{i+1}\mid\underline{h}_{i+1},\underline{h}_0\right)\cdot g_{\underline{H}_{i+1}|\underline{H}_0}\left(\underline{h}_{i+1}\mid\underline{h}_0\right)}{\kappa_{\underline{S}_{i+1}|\underline{H}_0}\left(\underline{s}_{i+1}\mid\underline{h}_0\right)}$$

$$(5.2)$$

where

$$\kappa_{\underline{S}_{i+1}|\underline{H}_0}\left(\underline{s}_{i+1}\mid\underline{h}_0\right)=\int_0^{+\infty}\int_0^{+\infty}\cdots\int_0^{+\infty}f_{\underline{S}_{i+1}|\underline{H}_{i+1},\underline{H}_0}\left(\underline{s}_{i+1}\mid\underline{h}_{i+1},\underline{h}_0\right)\cdot g_{\underline{H}_{i+1}|\underline{H}_0}\left(\underline{h}_{i+1}\mid\underline{h}_0\right)d\underline{h}_{i+1}$$

$$(5.3)$$

Similar to Eq. (5.1), as $\underline{H}_{i+1}$ is a vector, $\kappa_{\underline{S}_{i+1}|\underline{H}_0}\left(\underline{s}_{i+1}|\underline{h}_0\right)$ has to be evaluated by resolving a multiple integral. The conditional multivariate distributions related to Eq. (5.2) can be modelled by using one or more approaches suggested in Chapter 3.

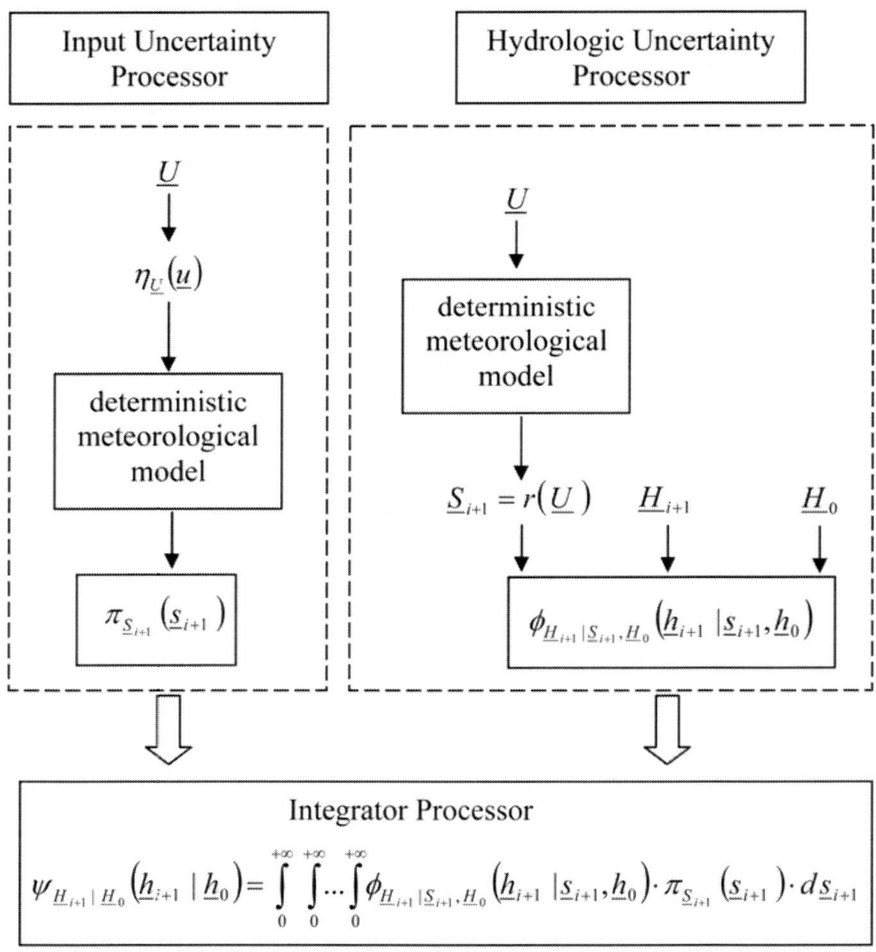

Figure 5.1. Structure of a BFS for coupling stochastic and meteorological models.

5.2. EXAMPLE OF A COUPLED MODEL: PRAISE-ME

According to the theory mentioned in Section 5.1, a temporal coupled model was developed at the University of Calabria, named Prediction of Rainfall Amount Inside Storm Events with MEteo or PRAISE-ME (De Luca 2006; De Luca and Versace 2010; De Luca et al. 2010). In PRAISE-ME, the meteorological model MM5, described in Chapter 4, is considered a NWP model, and the PRAISE model is adopted as the prior distribution. This coupled model was applied to a 9 km square cell in which the Cosenza rain gauge (Figure 3.2) was inside. For the calibration, observed hourly rainfall heights from the rain gauge as well as the database of predicted hourly rainfall by MM5 are considered.

The following sections describe the calibration of the coupled model (Section 5.2.1), and the validation results (Section 5.2.2)

5.2.1. PRAISE-ME Calibration

In a temporal model, the random vectors $\underline{H}_{i+1}$ and $\underline{S}_{i+1}$ become scalar random variables H_{i+1} and S_{i+1}. Only $\underline{H}_0$ remains a vector quantity, as it can be constituted by several antecedent rainfall heights before beginning the forecast.

Focusing on PRAISE-ME, the prior density is the conditional probability density of the PRAISE Model:

$$g_{H_{i+1}|\underline{H}_0}\left(h_{i+1}|\underline{h}_0\right) = f_{H_{i+1}|Z_i^{(v)}}\left(h_{i+1} \mid z_i^{(v)}\right) \tag{5.4}$$

According to Section 3.2.1.4 (Figure 3.1), it must be highlighted that with t_0 as the current time, the array $\underline{h}_0$ in the first temporal forecasting period $[t_0; t_0+\Delta t]$ is derived using the measured precipitation within v antecedent temporal interval. For the successive periods, $[t_0+(i-1)\Delta t; t_0+i\Delta t]$ (i.e. $i > 1$), the vector $\underline{h}_0$ is derived using $(v-i)$ observed rainfall values and i forecasted values from the PRAISE model.

As suggested by Krzysztofowicz (1999) and Krzysztofowicz and Herr (2001), the likelihood function was evaluated using the sample of joint observations $\left\{ \left(s_{i+1}^{(j)}, h_{i+1}^{(j)}, \left(z_i^{(v)} \right)^{(j)} \right), j = 1,..n \right\}$ related to Cosenza, in which:

1. s_{i+1} is the hourly rainfall forecasted by MM5 into the 9 km square pixel containing the rain gauge;

2. h_{i+1} is the measured hourly rainfall height at the rain gauge;

3. $z_i^{(v)}$ is the value of the linear function referred to the antecedent precipitations.

The random variables H_{i+1}, S_{i+1} and $Z_i^{(v)}$ are characterized by a mixed nature, and thus an expression similar to Eq. (3.37) was developed. In particular, in an analogy with the PRAISE and PRAISEST model, all the univariate distributions were modelled by adopting Weibull functions. For bivariate and trivariate cases, power transformations of Moran-Dowton exponential distributions were considered (see Section 3.2.3). Obviously, this choice does not exclude the possibility of adopting other distributions, e.g. Meta-Gaussian or Copula approaches (see Section 3.2.4-3.2.5).

In the following equations, for notation simplicity, the subscripts and the superscripts of the random variables are removed.

In this context, the posterior density is:

$$\phi_{H|S,Z}(h|\,s,\,z\,) = \frac{f_{S|H,Z}(s|\,h,z)\cdot f_{H|Z}(h|\,z)}{\kappa_{S|Z}(s|\,z)} \tag{5.5}$$

with $\kappa_{S|Z}(s|z) = \displaystyle\int_0^{+\infty} f_{S|H,Z}(s|h,z)\cdot f_{H|Z}(h|z)\,dh$.

Parameter evaluation was carried out by using the procedure reported in Section 3.2.3. In detail, the likelihood function was calibrated using the data within the period 2000-2004 (i.e. contemporaneous hourly measured data and MM5 forecasted data). The prior density in the PRAISE model was calibrated using the data within the period 1990-2004. Tables 1 and 2 contain the obtained calibration results for both the prior and likelihood functions. As a means to show the goodness-of-fit of the adopted distributions, Weibull probabilistic papers were used (see Appendix B for further details). As an example, Figure 5.2 shows the comparisons between empirical and theoretical

CDFs, for the variables (a) $H\,|\,H>0\cap S>0\cap Z>0$, (b) $S\,|\,H>0\cap S>0\cap Z>0$, and (c) $Z\,|\,H>0\cap S>0\cap Z>0$.

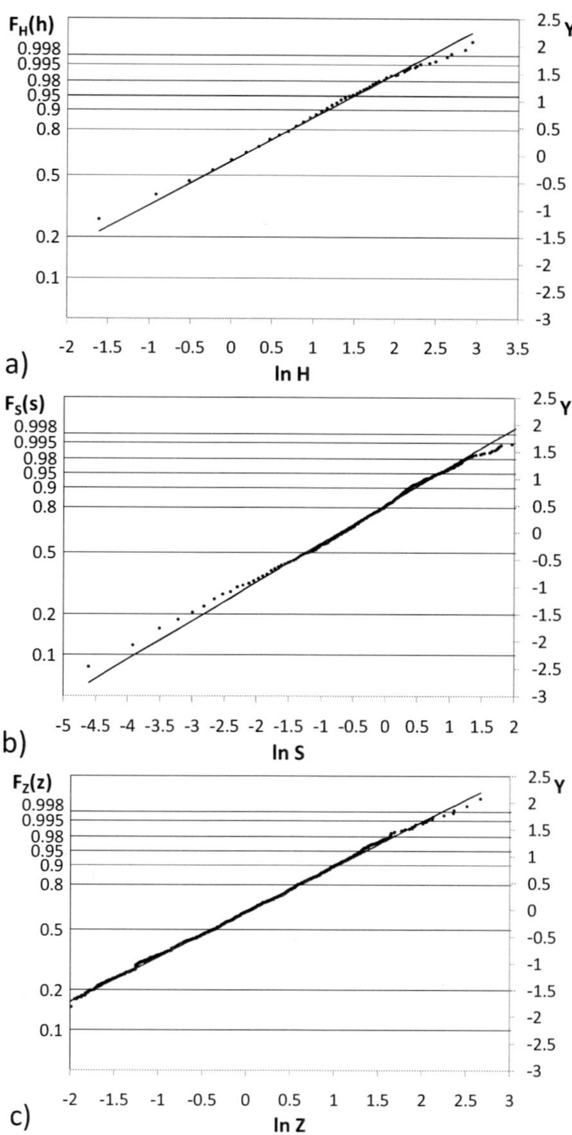

Figure 5.2. Likelihood function for Cosenza rain gauge: plots on Weibull probabilistic papers of empirical and Weibull distributions for the variables (a) $H\,|\,H>0\cap S>0\cap Z>0$ (b) $S\,|\,H>0\cap S>0\cap Z>0$ and (c) $Z\,|\,H>0\cap S>0\cap Z>0$.

Table 5.1. Cosenza rain gauge: calibration results for the prior density function (PRAISE model)

event	p	$1/\lambda_h$	η_h	$1/\lambda_z$	η_z	θ
	(-)	(mm)	(-)	(mm)	(-)	(-)
$H > 0 \cap Z > 0$	0.09	1.04	0.80	0.90	0.80	1.63
$H > 0 \cap Z = 0$	0.01	0.72	0.63			
$H = 0 \cap Z > 0$	0.14			0.31	0.52	

Table 5.2. Cosenza rain gauge: calibration results for the likelihood function

event	p	$1/\lambda_h$	η_h	$1/\lambda_s$	η_s	$1/\lambda_z$	η_z	θ
	(-)	(mm)	(-)	(mm)	(-)	(mm)	(-)	(-)
$H > 0 \cap S > 0 \cap Z > 0$	0.06	1.32	0.95	0.57	0.48	1.08	0.90	1.81
$H > 0 \cap S > 0 \cap Z = 0$	0.01	0.80	0.64	0.51	0.64			1.11
$H > 0 \cap S = 0 \cap Z > 0$	0.03	0.91	0.74			0.79	0.75	1.70
$H = 0 \cap S > 0 \cap Z > 0$	0.06			0.47	0.49	0.35	0.53	1.65
$H > 0 \cap S = 0 \cap Z = 0$	0.01	0.65	0.70					
$H = 0 \cap S > 0 \cap Z = 0$	0.11			0.35	0.50			
$H = 0 \cap S = 0 \cap Z > 0$	0.09					0.28	0.53	

5.2.2. PRAISE-ME Validation

First, the validation was carried out focusing attention on events "outside" the storm development. They are characterized by $Z_0 = 0$ mm (where the subscript "0" is related the eight previous hours of observed data) and by a positive meteorological forecasted value aggregated on the successive six hours. Both PRAISE and PRAISE-ME performances were tested on reproducing, for each of the successive six hours of forecasting, the following sample moments:

- mean $m_{H|Z_0=0}$;

- standard deviations $s_{H|Z_0=0}$;

- autocorrelation of lag 1 $r_{1|Z_0=0}$.

The results are shown in Figure 5.3, in which the bars represent the sample values for Cosenza rain gauge. Moreover for each index, mean value and 95% confidence intervals of PRAISE and PRAISE-ME simulations are shown.

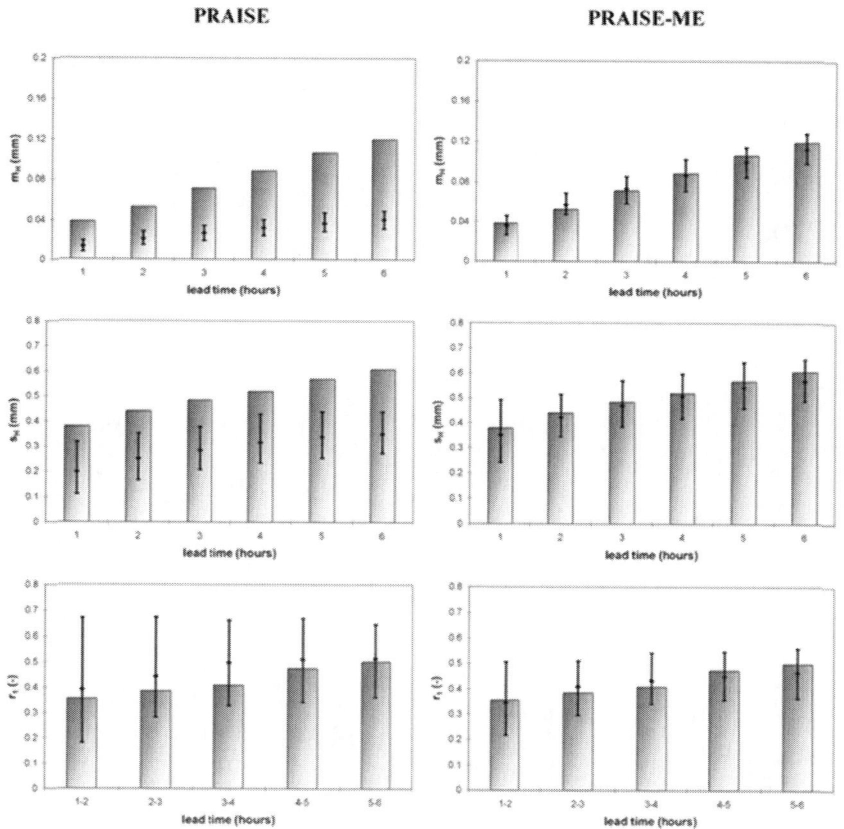

Figure 5.3. Validation results referred to PRAISE and PRAISE-ME for mean value $m_{H|Z_0=0}$, standard deviation value $s_{H|Z_0=0}$ and autocorrelation of lag 1 $r_{1|Z_0=0}$.

The overall capacity of the PRAISE-ME posterior density to reproduce the rainfall field properties is much higher than that of the PRAISE prior density. Consequently, for these events, the uncertainty regarding rainfall nowcasting can be significantly reduced by using the coupled meteo-stochastic model. In particular, for $m_{H|Z_0=0}$ and $s_{H|Z_0=0}$, the bias (i.e. the difference between the sample value of the quantity of interest and the mean value derived from the simulations) decreases greatly, tending to unbiased evaluations. For $r_{1|Z_0=0}$, both the variability and bias are reduced. The set of events "inside" the storm development characterized by $Z_0 > 1$ mm, either for a positive meteorological prediction or for a null one, was also considered. In this case, PRAISE and PRAISE-ME capabilities to reproduce rainfall properties are approximately the same.

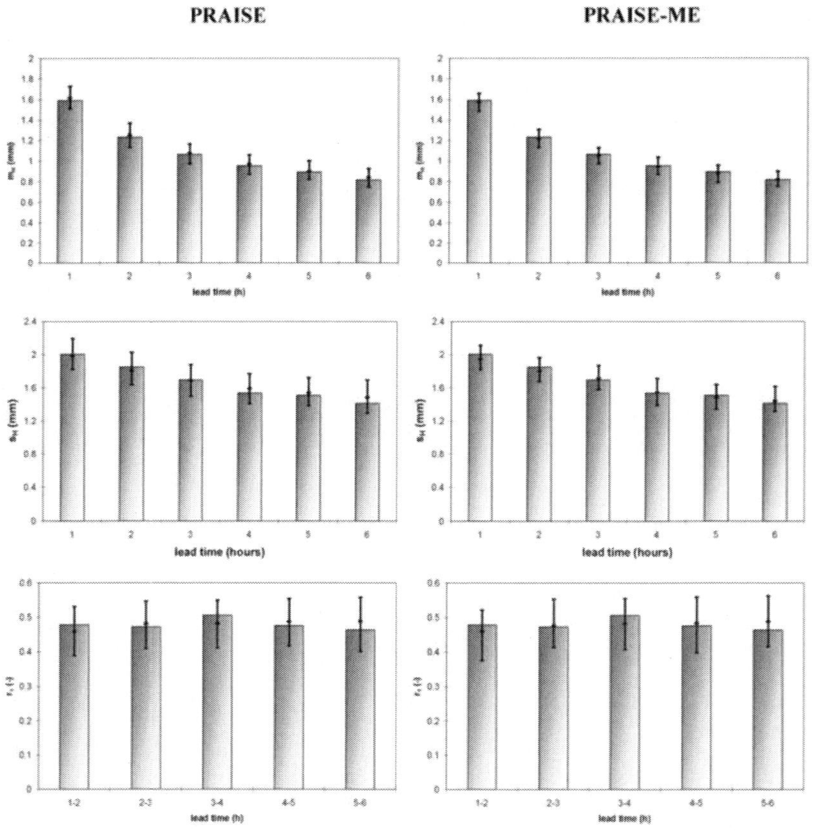

Figure 5.4. Validation results referred to PRAISE and PRAISE-ME for mean value $m_{H|Z_0>1}$, standard deviation value $s_{H|Z_0>1}$ and autocorrelation of lag 1 $r_{1|Z_0>1}$.

Figure 5.4 shows the following sample moments and for each hour of nowcasting:

- mean $m_{H|Z_0>1}$;

- standard deviations $s_{H|Z_0>1}$;

- autocorrelation of lag 1 $r_{1|Z_0>1}$.

This is justified by the PRAISE characteristics. In the case of nowcasting "inside a storm event", its prediction already constitutes a good reconstruction of future precipitations. Hence, no significant uncertainty reduction could be derived by using the PRAISE-ME model.

Moreover, as examples of model output, the applications to the December 6, 2002 and February 4, 2003 events for Cosenza rain gauge are shown in Figures 5.5-5.6.

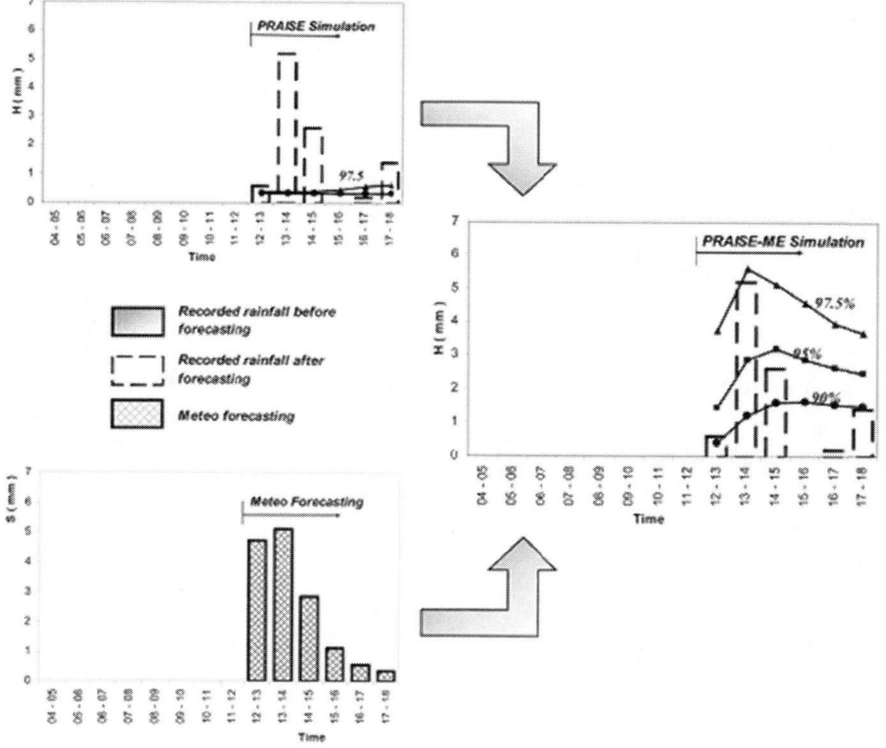

Figure 5.5. Applications of PRAISE and PRAISE-ME related to the event of December 6, 2002.

For both events, the observed rainfall values have been compared with the percentiles 90%, 95% and 97.5% of the prior and posterior distributions, for each forecasting hour. The rainfall events, here illustrated, present the following features:

- For the December 6, 2002 event, the memory extension of the observed rainfall starts from 4:00 LT to 12:00 LT and it is characterized by null precipitation. The forecasting period starts from 12:00 LT to 18:00 LT, and there is a positive meteorological prediction aggregate in this lapse time.
- For the February 4, 2003 event, the memory extension of the observed rainfall starts from 14:00 LT to 22:00 LT and it is characterized by $Z_0 > 1$ mm. The forecasting period starts from 22:00 LT to 04:00 LT on February 5th, and there is a positive meteorological prediction aggregate in this lapse time.

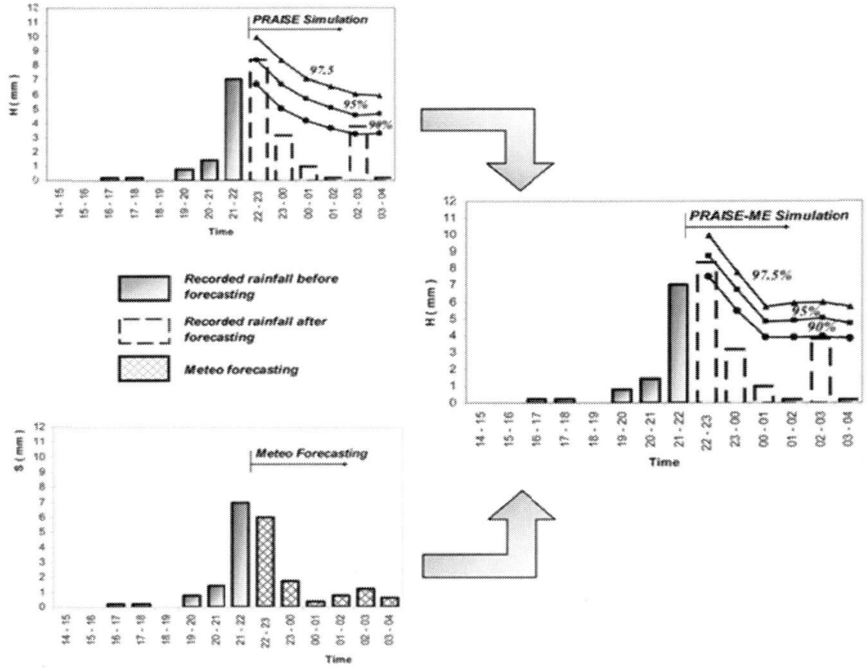

Figure 5.6. Applications of PRAISE and PRAISE-ME related to the event of February 4, 2003.

For the December 6, 2002 event, Figure 5.5 clearly shows that the forecasts using PRAISE-ME are superior to those of PRAISE. The PRAISE-ME forecasts are able to produce statistical confidence intervals, which contain the observed rainfall heights in the forecasting period.

On the other hand, the forecasts related to the February 4, 2003 event (Figure 5.6) indicate that there is no significant difference between PRAISE and PRAISE-ME predictions. The forecast is carried out "inside a storm event", and consequently each correspondent percentile has the same capability to contain or not to contain the observed rainfall.

CONCLUSION

This chapter highlighted the importance of a coupled modelling as a means to improve the nowcasting for "outside a storm event". On the other hand, as pointed out several times in this book, the use of stochastic models is sufficient for nowcasting "inside a storm event".

The model developed at the University of Calabria, and described in Section 5.2, constitutes a simple example relating to temporal modelling.

If the spatial scale of the study area is relevant, for which the hypothesis of uniformly distributed precipitation cannot be assumed, a more general coupling can be developed, by adopting the methodology reported in Section 5.1.

Chapter 6

CONCLUSION

Short term forecasting or nowcasting of rainfall is a crucial component in an early warning system. A rainfall prediction in real time is important in cases where rainfall induced phenomena can evolve within a very short time, such as flash floods and shallow landslides. Further, warnings based on only measured rainfall are not good, as they do not provide sufficient time for the activation of civil protection measures. As such, rainfall nowcasting is necessary. This book provides information for both developers and end-users of rainfall nowcasting models.

First, attention must be focused on the importance of a Probabilistic Quantitative Precipitation Forecast or PQPF (Chapter 2), which offers benefits with respect to a Quantitative Precipitation Forecast (QPF). Unlike a QPF, a PQPF allows the ability to quantify the uncertainty related to a forecast. Thus, if rainfall fields in terms of PQPF are the input for rainfall-runoff models or models for landslides triggered by rainfall, it is possible to evaluate in real time the probability of occurrence concerning all hydrological induced phenomena. An authority is therefore able to set risk-based criteria for watches, warnings and emergency responses.

In the technical literature, several methodologies are suggested for rainfall nowcasting. In the case of nowcasting "inside a storm event", stochastic models are reported to be more suitable for applications at hydrological spatial and temporal scales with respect to meteorological models. Chapter 3 provides an overview of many types of stochastic models, which are grouped into a model driven class (ARMA and models based on mixed distributions) and a data driven class (like ANNs). For the models within the model driven class, it is possible to obtain the exceedence or non-exceedence probability of given threshold values, or forecasts of particular exceedence or non-exceedence percentiles. For the models within the data driven class, a typical framework for QPF is proposed in literature. It is actually also possible to develop a Bayesian approach for probabilistic analysis (Section 3.3.3)

A brief overview of meteorological models, also named as Numerical Weather Prediction (NWP), is reported in Chapter 4. The overview is a qualitative discussion of the governing equations, classification, error sources and ensemble prediction systems.

Chapter 5 reports the methodology of coupling meteorological and stochastic models. This coupling enables improvement of the rainfall prediction at a hydrological scale for cases of nowcasting "outside a storm event".

In conclusion, to develop rainfall nowcasting modules for an early warning system, the following guidelines are suggested.

1. The choice of a temporal or spatiotemporal PQPF depending on the spatial scale of interest in the early warning system. The temporal PQPF is suitable for small areas where uniformly distributed precipitation can be assumed. The spatiotemporal PQPF is recommended for medium to large zones and for modeling on a regional scale.
2. The selection of different approaches (i.e. model driven or data driven), aimed at evaluation of "the best" model by comparing the model performance in the validation phase.
3. The use of a Bayesian structure for coupling NWP and stochastic models. It is particularly important for the cases where, in real time, the antecedent precipitation is null and a rainfall event may occur. On the other hand, in the case of nowcasting "inside a storm event", the use of a purely stochastic model is sufficient for obtaining reliable predictions.

APPENDIX A. PROBABILITY DISTRIBUTIONS

This appendix provides a brief overview related to the theory of probability distributions.

In detail, Section A.1 reports the basic concepts of probability, Section A.2 explain the type of random variables and associated functions, Section A.3 describes how to evaluate the fast characteristics (theoretical moments), and Section A.4 contains a list of probability distributions.

A.1. BASIC CONCEPTS OF PROBABILITY

First, the followings are defined:

1. Random experiment is an experiment that can result in different outcomes, even though it is repeated in the same manner every time.
2. Sample space is the set of all possible outcomes, which can occur in a random experiment.
3. Event is a subset of the sample space of a random experiment.

Then, by using Algebra of Events, several relations can be carried out among events named A, B, C, D and E:

- Intersection $C = A \cap B$: C occurs if both A and B occur;
- Union $D = A \cup B$: D occurs if either A or B or both occur;
- Complement $E = \overline{A}$: E occurs if A does not occur.

Moreover, the following are defined:

- Certain event I is an event equals to the sample space, thus it always occurs.
- Impossible event $\varnothing$ is an event that never occurs.
- Mutually exclusive events are events that they cannot occur in the same experiment, i.e. if events A and B are mutually exclusive then $A \cap B = \varnothing$
- Collectively exhaustive events are events for which the union is equal to the certain event I, i.e. events A and B are collectively exhaustive if $A \cup B = I$.

In this context, probability is a likelihood that an event will occur under a specified set of conditions. The principles that control probability are randomness and chance variations. Statistics deal with the collection, presentation, analysis and use of data to make decisions and solve problems. Statistical techniques (see Appendix B) can be used to describe and understand variability.

A.1.1. Axioms of Probability

Let $P[A]$ be the probability associated to an event A; the theory of probability is characterized by the following axioms:

1. $P[A] \geq 0$;
2. $P[I] = 1$;
3. If $A \cap B = \varnothing$ then $P[A \cup B] = P[A] + P[B]$;
4. Probability of B if previously A occurred, i.e. $P[B \mid A]$ (named conditional probability), with $P[A] \neq 0$, is equal to:

$$P[B \mid A] = \frac{P[A \cap B]}{P[A]} \tag{A.1}$$

The last axiom defines the relationship of dependence between a couple of events. Consequently, A and B are independent if $P[B]$ does not change whether or not A occurs and vice versa, i.e.:

$$P[B \mid A] = P[B] \tag{A.2}$$

From Eq. (A.2), for independent events, the following property is obtained:

$$P[B \mid A] = P[B] = \frac{P[A \cap B]}{P[A]} \Rightarrow P[A \cap B] = P[A]P[B] \tag{A.3}$$

A.1.2. Theorems of Probability

The following theorems are obtained by the axioms:

1. $P[\varnothing] = 0$
2. $P[\overline{A}] = 1 - P[A]$
3. $P[\overline{A} \cap B] = P[B] - P[A \cap B]$
4. $P[A \cup B] = P[A] + P[B] - P[A \cap B]$
5. Total Probability Theorem: let $\{B_1, B_2, ..., B_n\}$ be a set of mutually exclusive and collectively exhaustive events and let A be any other event. Then $P[A]$ can be obtained as:

$$P[A] = \sum_{i=1}^{n} P[A \cap B_i] = \sum_{i=1}^{n} P[B_i]P[A \mid B_i] \tag{A.4}$$

6. Bayes' Theorem:

$$P[A \mid B] = P[A]\frac{P[B \mid A]}{P[B]} \tag{A.5}$$

Using Total Probability Theorem, $P[B]$ can be expressed in terms of $P[A]$, $P[\overline{A}]$, and the conditional probabilities $P[B \mid A]$ and $P[B \mid \overline{A}]$:

$$P[B] = P[A]P[B \mid A] + P[\overline{A}]P[B \mid \overline{A}] \tag{A.6}$$

Thus Bayes' Theorem can be rewritten as:

$$P[A \mid B] = P[A] \frac{P[B \mid A]}{P[A]P[B \mid A] + P[\overline{A}]P[B \mid \overline{A}]} \tag{A.7}$$

A.2. Type of Random Variables and Properties of PDFs and CDFs

A random variable is a function that assigns real values to the outcomes of experiments. In detail, two types of random variables exist:

- Discrete random variables which are variables that can only assume discrete values (for example, the number of rainfall events occurring in one year).
- Continuous random variables are variables that can assume any value in a given range (for example, rainfall intensity during a storm event).

A random variable is denoted with capital letters (e.g. X, Y, and so on). A realization of a random variable (i.e. the event) is denoted with small letters (e.g. x, y, and so on).

A.2.1. Discrete Random Variables

For a discrete random variable X, it is possible to define a probability density function (PDF) as follows:

$$f_X(x_i) = P[X = x_i] \quad i = 1, 2, ..., k \tag{A.8}$$

in which k is the number of values which can be assumed by X, and k can be infinite. The PDF has the following properties:

1. $f_X(x_i) \geq 0$ $\tag{A.9}$

2. $\displaystyle\sum_{i=1}^{k} f_X(x_i) = 1$ (A.10)

3.

Moreover, the Cumulative Probability Function (CDF) is defined as:

$$F_X(x_i) = P[X \le x_i]$$ (A.11)

Its properties are:

1. $\displaystyle F_X(x_i) = \sum_{j=1}^{i} f_X(x_j)$ (A.12)

2. $\displaystyle \lim_{x_i \to -\infty} F_X(x_i) = 0 \; ; \; \lim_{x_i \to +\infty} F_X(x_i) = 1 \; ;$ (A.13)

3. *if $a < b$ then* $F_X(a) < F_X(b)$ (A.14)
    i.e. $F_X(.)$ is monotonic non-decreasing.

4. $P[a \le X \le b] = F_X(b) - F_X(a)$ (A.15)

A.2.2. Continuous Random Variables

A continuous random variable X can assume any value in a given range (between $-\infty$ and $+\infty$). Unlike a discrete random variable, it can assume an infinite and uncountable set of values. Thus, it is impossible to evaluate the probability for an assigned value x. Consequently, the PDF is defined as:

$$f_X(x)dx = P[x \le X \le x + dx]$$ (A.16)

Unlike Section A.2.1., $f_X(x)$ is related to the probability for X to fall inside an infinitesimal interval dx of a value x. It is characterized by the following properties:

1. $f_X(x) \ge 0 \; \text{-}\infty < x < +\infty$ (A.17)

2. $\displaystyle\int_{-\infty}^{+\infty} f_X(x)dx = 1$ (A.18)

Also in this case, a CDF is defined with the following properties:

1. $\displaystyle F_X(x) = P[X \le x] = \int_{-\infty}^{x} f_X(x)dx$ (A.19)

2. $\displaystyle \lim_{x \to -\infty} F_X(x) = 0 ; \ \lim_{x \to +\infty} F_X(x) = 1 ;$ (A.20)

3. $if \ a < b \ then \ F_X(a) < F_X(b)$ (A.21)
   i.e. $F_X(.)$ is monotonic non-decreasing.

4. $\displaystyle P[a \le X \le b] = F_X(b) - F_X(a) = \int_{a}^{b} f_X(x)dx$ (A.22)

From Eqs. (A.19-A.22), the following can be derived:

- $\displaystyle P[X = x_0] = \int_{x_0}^{x_0} f_X(x)dx = 0$ (A.23)

- $P[X \le x_0] = P[X < x_0]$ (A.24)

For each value of x in which $F_X(.)$ is differentiable, the following relationship holds

$$f_X(x) = \frac{dF_X(x)}{dx}$$ (A.25)

A.2.3. Overview on Multivariate Analysis

In the case where two or more random variables are considered in the statistical analysis, there is the need of defining multivariate PDFs and CDFs. In this section, only continuous random variables are examined. The treatment of discrete variables is quite straightforward.

Starting from the analysis of two random variables, a joint PDF can be obtained as:

$$f_{X_1,X_2}(x_1,x_2)dx_1dx_2 = P\left[x_1 \le X_1 \le x_1 + dx_1 \cap x_2 \le X_2 \le x_2 + dx_2\right]$$

(A.26)

in which

$$\int_{-\infty}^{+\infty}\int_{-\infty}^{+\infty}f_{X_1,X_2}(x_1,x_2)dx_1dx_2 = 1$$

(A.27)

The corresponding joint CDF is:

$$F_{X_1,X_2}(x_1,x_2) = P\left[X_1 \le x_1 \cap X_2 \le x_2\right] = \int_{-\infty}^{x_1}\int_{-\infty}^{x_2}f_{X_1,X_2}(x_1,x_2)dx_1dx_2$$

(A.28)

Similar to the univariate case (Section A.2.2), one obtains:

$$f_{X_1,X_2}(x_1,x_2) = \frac{\partial^2 F_{X_1,X_2}(x_1,x_2)}{\partial x_1 \partial x_2}$$

(A.29)

The marginal PDF of X_1 is defined as the distribution that is independent on the values of X_2. It is evaluated by integrating the joint PDF over all possible values of X_2, as follows:

$$f_{X_1}(x_1) = \int_{-\infty}^{+\infty}f_{X_1,X_2}(x_1,x_2)dx_2$$

(A.30)

The conditional CDF of X_1 given $X_2 = x_2$ is defined as:

$$F_{X_1|X_2}(x_1|x_2) = P\left[X_1 \le x_1 | X_2 = x_2\right]$$

(A.31)

The corresponding conditional PDF is obtained by the following differentiation:

$$f_{X_1|X_2}(x_1 \mid x_2) = \frac{dF_{X_1|X_2}(x_1 \mid x_2)}{dx_1} \tag{A.32}$$

with

$$\int_{-\infty}^{+\infty} f_{X_1|X_2}(x_1 \mid x_2)\,dx_1 = 1 \tag{A.33}$$

Referring to Eq. (A.1), the joint, conditional and marginal PDFs are related as follows:

$$f_{X_1,X_2}(x_1,x_2) = f_{X_1|X_2}(x_1 \mid x_2)f_{X_2}(x_2) = f_{X_2|X_1}(x_2 \mid x_1)f_{X_1}(x_1) \tag{A.34}$$

Moreover, it is possible to rewrite Bayes' Theorem as:

$$f_{X_1|X_2}(x_1 \mid x_2) = \frac{f_{X_1,X_2}(x_1,x_2)}{f_{X_2}(x_2)} = \frac{f_{X_2|X_1}(x_2 \mid x_1)f_{X_1}(x_1)}{\int_{-\infty}^{+\infty} f_{X_2|X_1}(x_2 \mid x_1)f_{X_1}(x_1)\,dx_1} \tag{A.35}$$

If X_1 and X_2 are independent, then

$$f_{X_1,X_2}(x_1,x_2) = f_{X_1}(x_1)f_{X_2}(x_2) \tag{A.36}$$

$$f_{X_1|X_2}(x_1 \mid x_2) = f_{X_1}(x_1) \tag{A.37}$$

Extension of Eq. (A.28) to the multidimensional case is straightforward. A M-dimensional joint CDF is defined as:

$$F_{\underline{X}}(\underline{x}) = F_{X_1,\dots,X_M}(x_1,\dots,x_M) = P\!\left[X_1 \le x_1 \cap \dots \cap X_M \le x_M\right] \tag{A.38}$$

and the corresponding PDF is evaluated with the following differentiation:

$$f_{\underline{X}}(\underline{x}) = f_{X_1,\dots,X_M}(x_1,\dots,x_M) = \frac{\partial^M F_{X_1,\dots,X_M}(x_1,\dots,x_M)}{\partial x_1 \cdots \partial x_M} \tag{A.39}$$

If all the random variables are independent, then:

$$f_{X_1,\ldots,X_M}(x_1,\ldots,x_M) = f_{X_1}(x_1)\cdots f_{X_M}(x_M) \tag{A.40}$$

Furthermore, if the random vector $\underline{X}$ is rewritten, in an equivalent way, as a partition, i.e.

$$\underline{X} = (\underline{X}_1, \underline{X}_2) \tag{A.41}$$

then Eq. (A.39) can be rewritten as:

$$f_{\underline{X}}(\underline{x}) = f_{\underline{X}_1,\underline{X}_2}(\underline{x}_1,\underline{x}_2) \tag{A.42}$$

Conditional PDFs can then be represented as:

$$f_{\underline{X}_1|\underline{X}_2}(\underline{x}_1\,|\,\underline{x}_2) = \frac{f_{\underline{X}_1,\underline{X}_2}(\underline{x}_1,\underline{x}_2)}{f_{\underline{X}_2}(\underline{x}_2)} = \frac{f_{\underline{X}_2|\underline{X}_1}(\underline{x}_2\,|\,\underline{x}_1)f_{\underline{X}_1}(\underline{x}_1)}{\int_{all\,\underline{x}_1} f_{\underline{X}_2|\underline{X}_1}(\underline{x}_2\,|\,\underline{x}_1)f_{\underline{X}_1}(\underline{x}_1)d\underline{x}_1} \tag{A.43}$$

A.3. FAST CHARACTERISTICS: THEORETICAL MOMENTS

While a random variable X can be fully explained by its PDF and CDF, it is nevertheless sometimes useful to use the fast characteristics (also known as the theoretical moments) particularly when different probability distributions need to be compared. The following is a summary of specific aspects of the fast characteristics.

A.3.1. Expected Value

The expected value μ_X of a random variable is the weighted average of all possible values that this random variable can assume. The weights used in computing this weighted average correspond to the probabilities in the case of a discrete random variable (see Eq. A.44), or the densities in the case of a continuous random variable (see Eq. A.45):

$$\mu_X = E[X] = \sum_{i=1}^{k} x_i f_X(x_i)$$
(A.44)

$$\mu_X = E[X] = \int_{-\infty}^{+\infty} x f_X(x) dx$$
(A.45)

$E[.]$ is the expectation operator, characterized by the following properties (a, b and c are constant values):

$$E[a] = a$$
(A.46)

$$E[aX] = aE[X]$$
(A.47)

$$E[X_1 + X_2] = E[X_1] + E[X_2]$$
(A.48)

$$E[a + bX_1 + cX_2] = a + bE[X_1] + cE[X_2]$$
(A.49)

A.3.2. Moments of Order R for a Random Variable X

Let $R \in N^+$ be the order, the moment of order R is defined as:

$$\mu_X^{(R)} = E[X^R]$$
(A.50)

i.e. it is the expected value of the random variable X^R.

For discrete random variables, the expression is:

$$\mu_X^{(R)} = \sum_{i=1}^{k} x_i^R f_X(x_i)$$
(A.51)

and for continuous random variables, the expression is:

$$\mu_X^{(R)} = \int_{-\infty}^{+\infty} x^R f_X(x) dx$$
(A.52)

It is apparent that $\mu_X = \mu_X^{(1)}$.

A.3.3. Central Moments of Order R with respect to E[X]

By defining the random variable Y as:

$$Y = (X - \mu_X)$$ (A.53)

the associated moment of order R is:

$$\overline{\mu_X^{(R)}} = E[Y^R] = E[(X - \mu_X)^R] = \sum_{i=1}^{k} (x_i - \mu_X)^R f_X(x_i)$$ (A.54)

for discrete random variables, and

$$\overline{\mu_X^{(R)}} = E[Y^R] = E[(X - \mu_X)^R] = \int_{-\infty}^{+\infty} (x - \mu_X)^R f_X(x)dx$$ (A.55)

for continuous random variables.

An important moment for Y is the variance, representing the dispersion of the X values with respect to μ_X :

$$\overline{\mu_X^{(2)}} = E[Y^2] = E[(X - \mu_X)^2] = \sigma_X^2 = Var[X] = \int_{-\infty}^{+\infty} (x - \mu_X)^2 f_X(x)dx$$

(A.55)

Where $\sigma_X = \sqrt{\sigma_X^2}$ is the defined standard deviation. The $Var[.]$ operator is characterized by the following properties (a is a constant value):

$$Var[X^2] = E[X^2] - \mu_X^2$$ (A.56)

$$Var[a] = 0$$ (A.57)

$$Var[X + a] = Var[X] \tag{A.58}$$

$$Var[aX] = a^2 Var[X] \tag{A.59}$$

$$Var[aX + b] = a^2 Var[X] \tag{A.60}$$

If $\mu_X \neq 0$, then it is possible to quantity the dispersion of X values with the variation coefficient V_X:

$$V_X = \frac{\sigma_X}{\mu_X} \tag{A.61}$$

A.3.4. Moments of Order R with Respect to the Standardized Random Variable Z

By defining the standardized random variable Z as:

$$Z = \frac{X - \mu_X}{\sigma_X} \tag{A.62}$$

The associated moment of order R is:

$$\overline{\mu_X^{(R)}} = E[Z^R] = E\left[\left(\frac{X - \mu_X}{\sigma_X}\right)^R\right] = \sum_{i=1}^{k}\left(\frac{x_i - \mu_X}{\sigma_X}\right)^R f_X(x_i) \tag{A.63}$$

for discrete random variables, and

$$\overline{\mu_X^{(R)}} = E[Z^R] = E\left[\left(\frac{X - \mu_X}{\sigma_X}\right)^R\right] = \int_{-\infty}^{+\infty}\left(\frac{x - \mu_X}{\sigma_X}\right)^R f_X(x)dx \tag{A.64}$$

for continuous random variables. It is possible to demonstrate that:

$$E[Z] = 0 \tag{A.65}$$

$$E\left[Z^2\right] = 1 \tag{A.66}$$

An important moment for Z is the skewness, which provides information about the symmetry/asymmetry of the PDF:

$$\gamma_1 = \overline{\overline{\mu_X^{(3)}}} = E\left[Z^3\right] = E\left[\left(\frac{X - \mu_X}{\sigma_X}\right)^3\right] \tag{A.67}$$

If $\gamma_1 = 0$ then the curve is symmetric. If $\gamma_1 > 0$ then the right tail of the curve is longer than the left tail. If $\gamma_1 < 0$ then the left tail of the curve is longer than the right tail.

Another important moment for Z is the kurtosis (or excess kurtosis):

$$\gamma_2 = \overline{\overline{\mu_X^{(4)}}} - 3 = E\left[Z^4\right] - 3 = E\left[\left(\frac{X - \mu_X}{\sigma_X}\right)^4\right] - 3 \tag{A.68}$$

It is a descriptor of the shape of a probability distribution. If $\gamma_2 > 0$ then the PDF is said to be leptokurtic, and has a more acute peak around the mean and fatter tails. If $\gamma_2 < 0$ then the PDF is said to be platykurtic, and has a lower, wider peak around the mean and thinner tails. The number three in Eq. (A.68) is the theoretical value of $E\left[Z^4\right]$ for a normal distribution (see Section A.4). If $\gamma_2 = 0$ then the PDF is said to be mesokurtic.

Table A.1 resumes the theoretical moments that can be evaluated for a random variable.

A.3.5. Other Fast Characteristics

Other common measures of the fast characteristics are:

- the median value $x_{0.5}$, for which $F_X(x_{0.5}) = 0.5$,
- the percentile (also named quantile) x_P, for which $F_X(x_P) = P$, and
- the mode x_{mod}, which is the value for which the PDF is maximum.

Table A.1. Theoretical moments for a random variable

Random variable	General formulation	Discrete random variable	Continuous random variable
X	$\mu_X^{(R)} = E\left[X^R\right]$ Expected value: $\mu_X = E[X]$	$\displaystyle\sum_{i=1}^{k} x_i^R f_X(x_i)$	$\displaystyle\int_{-\infty}^{+\infty} x^R f_X(x)dx$
$Y = (X - \mu_X)$	$\overline{\mu_X^{(R)}} = E\left[(X - \mu_X)^R\right]$ Variance: $\sigma_X^2 = E\left[(X - \mu_X)^2\right]$	$\displaystyle\sum_{i=1}^{k} (x_i - \mu_X)^R f_X(x_i)$	$\displaystyle\int_{-\infty}^{+\infty} (x - \mu_X)^R f_X(x)dx$
$Z = \dfrac{X - \mu}{\sigma}$	$\overline{\overline{\mu_X^{(R)}}} = E\left[\left(\dfrac{X - \mu_X}{\sigma_X}\right)^R\right]$ Skewness: $\overline{\overline{\mu_X^{(3)}}} = \gamma_1$ Kurtosis: $\overline{\overline{\mu_X^{(4)}}} - 3 = \gamma_2$	$\displaystyle\sum_{i=1}^{k} \left(\dfrac{x_i - \mu_X}{\sigma_X}\right)^R f_X(x_i)$	$\displaystyle\int_{-\infty}^{+\infty} \left(\dfrac{x - \mu_X}{\sigma_X}\right)^R f_X(x)dx$

A.3.6. Theoretical Moments for Multivariate Analysis

In the case of multivariate analysis, it is possible to evaluate the same theoretical moments as mentioned in Sections A.3.2-A.3.4, for each marginal and conditional PDF.

Moreover, for any couple of random variables X_i and X_j, two further measures can be obtained:

- the covariance $Cov[X_i, X_j]$:

$$Cov[X_i, X_j] = E[(X_1 - \mu_{X_1})(X_2 - \mu_{X_2})] \qquad (A.69)$$

If $X_i = X_j$ then $Cov[X_i, X_j] = \sigma_{X_i}^2$.

- the linear coefficient of correlation ρ_{X_i, X_j} :

$$\rho_{X_i, X_j} = \frac{Cov[X_i, X_j]}{\sigma_{X_i} \sigma_{X_j}} \qquad (A.70)$$

If $X_i = X_j$ then $\rho_{X_i, X_j} = 1$.

It must be pointed out that $Cov[X_i, X_j] = Cov[X_j, X_i]$ and $\rho_{X_i, X_j} = \rho_{X_j, X_i}$. The coefficient ρ_{X_i, X_j} is bounded by -1 and +1. If $\rho_{X_i, X_j} = \pm 1$ then X_i and X_j are perfectly correlated (positively or negatively). If $\rho_{X_i, X_j} = 0$ then X_i and X_j are uncorrelated. The independence between X_i and X_j implies $\rho_{X_i, X_j} = 0$, but not vice versa.

In general, by considering M random variables $X_1, ..., X_M$, two $M \times M$ symmetric matrices can be constructed: Covariance matrix $\underline{\underline{Cov}}$ and the correlation matrix $\underline{\underline{\rho}}$:

$$\underline{\underline{Cov}} = \begin{bmatrix} \sigma^2_{X_1} & Cov[X_1,X_2] & ... & Cov[X_1,X_M] \\ Cov[X_1,X_2] & \sigma^2_{X_2} & ... & ... \\ ... & ... & ... & ... \\ Cov[X_1,X_M] & ... & ... & \sigma^2_{X_M} \end{bmatrix} \qquad (A.71)$$

$$\underline{\underline{\rho}} = \begin{bmatrix} 1 & \rho_{X_1,X_2} & ... & \rho_{X_1,X_M} \\ \rho_{X_1,X_2} & 1 & ... & ... \\ ... & ... & ... & ... \\ \rho_{X_1,X_M} & ... & ... & 1 \end{bmatrix} \qquad (A.72)$$

A.4. LIST OF PROBABILITY DISTRIBUTIONS

In the following, a list of probability distributions is provided for both discrete and continuous random variables. For each distribution, details related to the characteristics of the random variable, PDF expression, parameters, theoretical moments (Expected value, Variance, Skewness and Kurtosis) are reported. The considered distributions are:

- Discrete random variables
  - Bernoulli distribution
  - Binomial distribution
  - Geometric distribution
  - Poisson distribution
  - Discrete Uniform distribution

- Continuous random variables
  - Continuous Uniform distribution
  - Normal distribution
  - Log-Normal distribution
  - Exponential distribution
  - Gamma distribution
  - Weibull Distribution

- o   Pareto distribution
- o   Log-Logistic distribution

A.4.1. Bernoulli Distribution

Characteristics of the random variable

A Bernoulli random variable X takes value 1 with success probability p and value 0 with failure probability $1-p$.

PDF

$$f_X(x) = p^x(1-p)^{1-x} = \begin{cases} p & \textit{if } X = 1 \\ 1-p & \textit{if } X = 0 \\ 0 & \textit{otherwise} \end{cases} \tag{A.73}$$

Parameters

p = success probability, with $0 \le p \le 1$.

Theoretical moments

$$\mu_X = p \tag{A.74}$$

$$\sigma_X^2 = p(1-p) \tag{A.75}$$

$$\gamma_1 = \frac{1-2p}{\sqrt{(1-p)p}} \tag{A.76}$$

$$\gamma_2 = \frac{1-6p(1-p)}{\sqrt{(1-p)p}} \tag{A.77}$$

A.4.2. Binomial Distribution

Characteristics of the random variable

A Binomial random variable X corresponds to the number x of successes out of Bernoulli n experiments, which are independent among them and characterized by the same success probability p . It is clear that $0 \le x \le n$.

PDF

$$f_X(x) = \binom{n}{x} p^x (1-p)^{n-x}$$ (A.78)

where $\binom{n}{x}$ is the binomial coefficient, defined as:

$$\binom{n}{x} = \frac{n!}{x!(n-x)!}$$ (A.79)

Parameters

p = success probability of each Bernoulli experiment, with $0 \le p \le 1$.

n = number of Bernoulli experiments, with $n > 0$.

Theoretical moments

$$\mu_X = np$$ (A.80)

$$\sigma_X^2 = np(1-p)$$ (A.81)

$$\gamma_1 = \frac{1-2p}{\sqrt{n(1-p)p}}$$ (A.82)

$$\gamma_2 = \frac{1-6p(1-p)}{\sqrt{n(1-p)p}}$$ (A.83)

A.4.3. Geometric Distribution

Characteristics of the random variable

A Geometric random variable X corresponds to the number x of Bernoulli experiments needed to get the first success. It is clear that $x > 0$.

PDF

$$f_X(x) = (1-p)^{x-1} p \qquad (A.84)$$

Parameters

p = success probability of each Bernoulli experiment, with $0 \le p \le 1$.

Theoretical moments

$$\mu_X = \frac{1}{p} \qquad (A.85)$$

$$\sigma_X^2 = \frac{(1-p)}{p^2} \qquad (A.86)$$

$$\gamma_1 = \frac{2-p}{\sqrt{1-p}} \qquad (A.87)$$

$$\gamma_2 = 6 + \frac{p^2}{1-p} \qquad (A.88)$$

A.4.4. Poisson Distribution

Characteristics of the random variable

A Poisson random variable X corresponds to the number x of successes that occur in an assigned interval of time and/or space. It is possible to demonstrate that the Poisson distribution is a particular case of the Binomial distribution when $n \to +\infty$ and $p \to 0$. Consequently, a Poisson random variable is suitable to model rare events.

PDF

$$f_X(x) = \frac{v^x e^{-v}}{x!} \qquad (A.89)$$

Parameters

v = average value of successes in an assigned interval of time and/or space. v is equal to the product np of Binomial parameters. By assuming $v = \lambda t$, Eq. (A.89) can be rewritten as:

$$f_X(x) = \frac{(\lambda t)^x e^{-\lambda t}}{x!} \qquad \text{(A.90)}$$

where λ is the average rate of successes and t is the assigned interval of time.

Theoretical moments

$$\mu_X = v \qquad \text{(A.91)}$$

$$\sigma_X^2 = v \qquad \text{(A.92)}$$

$$\gamma_1 = \frac{1}{\sqrt{v}} \qquad \text{(A.93)}$$

$$\gamma_2 = \frac{1}{v} \qquad \text{(A.94)}$$

A.4.5. Discrete Uniform Distribution

Characteristics of the random variable

A Discrete Uniform variable X assumes the same probability for each outcome x_i, $i = 1, 2, ..., N$.

PDF

$$f_X(x_i) = \begin{cases} 1/N & i = 1, 2, ..., N \\ 0 & \textit{otherwise} \end{cases} \qquad \text{(A.95)}$$

Parameters

N = number of values which can be assumed by the random variable

Theoretical moments

$$\mu_X = \frac{N+1}{2} \tag{A.96}$$

$$\sigma_X^2 = \frac{\left(N^2-1\right)}{12} \tag{A.97}$$

$$\gamma_1 = 0 \tag{A.98}$$

$$\gamma_2 = -\frac{6\left(N^2+1\right)}{5\left(N^2-1\right)} \tag{A.99}$$

A.4.6. Continuous Uniform Distribution

Characteristics of the random variable

A Continuous Uniform variable X assumes the same probability for each outcome $x \in [a,b]$.

PDF

$$f_X(x) = \begin{cases} \dfrac{1}{b-a} & \text{if } a \leq x \leq b \\[2mm] 0 & \text{otherwise} \end{cases} \tag{A.100}$$

Parameters

a = the lowest value which can be assumed by the random variable.
b = the highest value which can be assumed by the random variable.

Theoretical moments

$$\mu_X = \frac{a+b}{2} \tag{A.101}$$

$$\sigma_X^2 = \frac{(b-a)^2}{12} \tag{A.102}$$

$$\gamma_1 = 0 \tag{A.103}$$

$$\gamma_2 = -\frac{6}{5} \tag{A.104}$$

A.4.7. Normal Distribution

Characteristics of the random variable

A Normal variable X assumes values between $-\infty$ and $+\infty$ and it is characterized by a bell-shaped PDF (Eq. A.105)

PDF

$$f_X(x) = \frac{1}{\sigma\sqrt{2\pi}} e^{-\frac{(x-\mu)^2}{2\sigma^2}} \tag{A.105}$$

with $x \in (-\infty; +\infty), \mu \in (-\infty; +\infty), \sigma > 0$.

Parameters

μ = the expected value of the random variable. It also corresponds to modal and median value.
σ = the standard deviation value of the random variable.

Theoretical moments

$$\mu_X = \mu \tag{A.106}$$

$$\sigma_X^2 = \sigma^2 \tag{A.107}$$

$$\gamma_1 = 0 \tag{A.108}$$

$$\gamma_2 = 0 \tag{A.109}$$

If the standardized random variable Z is considered (Eq. A.62) then the standard normal distribution is obtained:

$$f_Z(z) = \frac{1}{\sqrt{2\pi}} e^{-\frac{z^2}{2}}$$

(A.110)

in which $\mu = 0$ and $\sigma^2 = 1$.

A.4.8. Log-Normal distribution

Characteristics of the random variable

A random variable X is Log-Normal if $Y = \ln X$ is a Normal random variable.

PDF

$$f_X(x) = \frac{1}{\sigma_Y\, x \sqrt{2\pi}} e^{-\frac{(\ln x - \mu_Y)^2}{2\sigma_Y^2}}$$

(A.111)

with $x > 0, \mu_Y \in (-\infty; +\infty), \sigma_Y > 0$

Parameters

μ_Y = the expected value of the Normal random variable $Y = \ln X$.

σ_Y = the standard deviation value of the Normal random variable $Y = \ln X$.

Theoretical moments

$$\mu_X = e^{\mu_Y + \frac{1}{2}\sigma_Y^2}$$

(A.112)

$$\sigma_X^2 = e^{2\mu_Y + 2\sigma_Y^2} - e^{\mu_Y + \sigma_Y^2}$$

(A.113)

$$\gamma_1 = \left(e^{\sigma_Y^2} + 2\right)\sqrt{e^{\sigma_Y^2} - 1}$$

(A.114)

$$\gamma_2 = e^{4\sigma_Y^2} + 2e^{3\sigma_Y^2} + 3e^{2\sigma_Y^2} - 6$$

(A.115)

A.4.9. Exponential Distribution

Characteristics of the random variable

An exponential random variable X is particularly suitable to model the time interval between two consecutive Poisson occurrences.

PDF

$$f_X(x) = \lambda e^{-\lambda x} \qquad\qquad (A.116)$$

with $x > 0, \lambda > 0$.

Parameters

λ = inverse of the expected value related to the temporal interval between two consecutive Poisson occurrences. It corresponds to λ in Eq. (A.90).

Theoretical moments

$$\mu_X = \frac{1}{\lambda} \qquad\qquad (A.117)$$

$$\sigma_X^2 = \frac{1}{\lambda^2} \qquad\qquad (A.118)$$

$$\gamma_1 = 2 \qquad\qquad (A.119)$$

$$\gamma_2 = 6 \qquad\qquad (A.120)$$

A.4.10. Gamma Distribution

Characteristics of the random variable

A Gamma random variable X is particularly suitable to model the time interval in which there are α Poisson occurrences.

PDF

$$f_X(x) = \frac{x^{\alpha-1} e^{-\frac{x}{\beta}}}{\beta^\alpha \Gamma(\alpha)} \qquad (A.121)$$

where $\Gamma(.)$ is the Complete Gamma function (Abramovitz and Stegun 1970) defined as:

$$\Gamma(\alpha) = \int_0^{+\infty} t^{\alpha-1} e^{-t} dt \qquad (A.122)$$

$\Gamma(.)$ has the following properties:
1. $\Gamma(\alpha) = (\alpha - 1)\Gamma(\alpha - 1)$ $\qquad (A.123)$
2. If α is integer, then
    $\Gamma(\alpha) = (\alpha - 1)!$ $\qquad (A.124)$

Parameters

α = the average value of Poisson occurrences into the temporal interval x.
β = the average value of the temporal interval between two consecutive Poisson occurrences.

Theoretical moments

$$\mu_X = \alpha\beta \qquad (A.125)$$

$$\sigma_X^2 = \alpha\beta^2 \qquad (A.126)$$

$$\gamma_1 = \frac{2}{\sqrt{\alpha}} \qquad (A.127)$$

$$\gamma_2 = \frac{6}{\alpha} \tag{A.128}$$

A.4.11. Weibull Distribution

Characteristics of the random variable

A Weibull random variable X is obtained by an exponential random variable Y, if a power transformation is defined, i.e.:

$$Y = \alpha X^{\beta} \tag{A.129}$$

PDF

$$f_X(x) = \alpha\beta x^{\beta-1} e^{-\alpha x^{\beta}} \tag{A.130}$$

with $x > 0, \alpha > 0, \beta > 0$.

Parameters

α = the factor of scale.
β = the factor of shape.

Theoretical moments

$$\mu_X = \frac{\Gamma(1 + 1/\beta)}{\alpha} \tag{A.131}$$

$$\sigma_X^2 = \frac{1}{\alpha^{2/\beta}}\left[\Gamma(1 + 2/\beta) - \Gamma^2(1 + 1/\beta)\right] \tag{A.132}$$

$$\gamma_1 = \frac{\Gamma(1 + 3/\beta)/\alpha^3 - 3\mu_X\sigma_X^2 - \mu_X^3}{\sigma_X^3} \tag{A.133}$$

$$\gamma_2 = \frac{\Gamma(1+4/\beta)/\alpha^4 - 4\gamma_1\mu_X\sigma_X^3 - 6\mu_X^2\sigma_X^2 - \mu_X^4}{\sigma_X^4} - 3 \tag{A.134}$$

A.4.12. Pareto Distribution

Characteristics of the random variable

If Y is exponentially distributed with parameter α then the random variable

$$X = \beta e^{Y} \tag{A.135}$$

is Pareto distributed, where β is the minimum value assumed by X.

PDF

$$f_X(x) = \frac{\alpha\beta^\alpha}{x^{\alpha+1}} \tag{A.136}$$

with $x > \beta, \alpha > 0, \beta > 0$.

Parameters

α = the factor of shape.
β = the factor of scale.

Theoretical moments

$$\mu_X = \frac{\alpha\beta}{\alpha-1} \qquad \text{for } \alpha > 1 \tag{A.137}$$

$$\sigma_X^2 = \frac{\beta^2\alpha}{(\alpha-1)^2(\alpha-2)} \qquad \text{for } \alpha > 2 \tag{A.138}$$

$$\gamma_1 = \frac{2(1+\alpha)}{\alpha-3}\sqrt{\frac{\alpha-2}{\alpha}} \qquad \text{for } \alpha > 3 \qquad \text{(A.139)}$$

$$\gamma_2 = \frac{6(\alpha^3 + \alpha^2 - 6\alpha - 2)}{\alpha(\alpha-3)(\alpha-4)} \qquad \text{for } \alpha > 4 \qquad \text{(A.140)}$$

A.4.13. Log-Logistic Distribution

PDF

$$f_X(x) = \frac{(\beta/\alpha)(x/\alpha)^{\beta-1}}{\left[1 + (x/\alpha)^\beta\right]^2} \qquad \text{(A.141)}$$

with $x > \beta, \alpha > 0, \beta > 0$.

Parameters

α = the factor of scale.
β = the factor of shape.

Theoretical moments

$$\mu_X = \frac{\alpha\,\pi/\beta}{\sin(\pi/\beta)} \qquad \text{for } \beta > 1 \qquad \text{(A.142)}$$

$$\sigma_X^2 = \alpha^2\left(\frac{2\pi/\beta}{\sin(2\pi/\beta)} - \frac{(\pi/\beta)^2}{\sin^2(2\pi/\beta)}\right) \qquad \text{for } \beta > 2 \qquad \text{(A.143)}$$

$$\gamma_1 = \frac{3\dfrac{\pi/\beta}{\sin(3\pi/\beta)} - 6\dfrac{(\pi/\beta)^2}{\sin(2\pi/\beta)\sin(\pi/\beta)} + 2\dfrac{(\pi/\beta)^3}{\sin^3(\pi/\beta)}}{\left(2\dfrac{\pi/\beta}{\sin(2\pi/\beta)} - \dfrac{(\pi/\beta)^2}{\sin^2(\pi/\beta)}\right)^{3/2}}$$

for $\beta > 3$ (A.144)

$$\gamma_2 = \frac{4\dfrac{\pi/\beta}{\sin(4\pi/\beta)} - 12\dfrac{(\pi/\beta)^2}{\sin(3\pi/\beta)\sin(\pi/\beta)} + 12\dfrac{(\pi/\beta)^3}{\sin(2\pi/\beta)\sin^2(\pi/\beta)} - 3\dfrac{(\pi/\beta)^3}{\sin^4(\pi/\beta)}}{\left(2\dfrac{\pi/\beta}{\sin(2\pi/\beta)} - \dfrac{(\pi/\beta)^2}{\sin^2(\pi/\beta)}\right)^2} - 3$$

for $\beta > 4$ (A.145)

APPENDIX B. STATISTICAL INFERENCE

Statistical inference means drawing conclusions based on data. One context for inference is the parametric model, in which data are supposed to come from a certain probability distribution (in terms of PDF and CDF), which is characterized by a set of unknown parameters to be estimated. In this context, it is possible to define four characteristic steps:

1. sample data analysis (Section B.1);
2. hypothesis about probability distribution, derived from the previous step (see Appendix A for a set of available probability functions);
3. parameter estimation (Section B.2);
4. model validation (Section B.3).

Moreover, relating to multivariate analysis, Section B.4 shows how to model the relationship among two or more random variables by using regression analysis. Section B.5 describes tests that can be used to validate a Meta-Gaussian distribution (Section 3.2.4 of Chapter 3).

B.1. SAMPLE DATA ANALYSIS

Sample data analysis constitutes the first step of statistical inference and it can be carried out in both a graphical (Section B.1.1) and an analytical (Section B.1.2) way. Depending on the shape of these plots and on the analytical evaluations, a hypothesis about theoretical distribution can be assumed.

B.1.1. Graphical Representations

Graphical representations are sample analogous of PDF and CDF. In details we refer to histograms (Section B.1.1.1) and plotting position visualizations (Section B.1.1.2), respectively.

B.1.1.1. Histograms

Histograms provide a visual impression of data distribution. It consists of tabular frequencies, organized as adjacent rectangles related to discrete intervals of the random variable. In details, three kinds of histogram can be realized:

- absolute frequency histogram;
- relative frequency histogram;
- frequency density histogram.

Absolute Frequency Histogram

Sample data are grouped into k number of intervals. The k value can be established by using the Sturges' formula, as follows:

$$k = 1 + 3.3 \log_{10} n \tag{B.1}$$

where n is the number of sample data. Once evaluated k, the size Δx of each interval can be calculated as:

$$\Delta x = \frac{x_{\max} - x_{\min}}{k} \tag{B.2}$$

in which $x_{\min}$ and $x_{\max}$ are the minimum and the maximum sample values. The lower and higher bounds of each interval can be estimated as:

$$x_i^{(\inf)} = x_{\min} + (i-1)\Delta x \tag{B.3}$$

$$x_i^{(\sup)} = x_{\min} + i\Delta x \tag{B.4}$$

with $i = 1, 2, ... k$

Finally, the height of the ith rectangle is equal to the number of data points within the ith interval. This number is named absolute frequency. As such, the summation of all the absolute frequencies must equal the number of data points.

Relative Frequency Histogram

An absolute frequency histogram can be normalized with respect to the total number of data points, i.e. the height of each rectangle (absolute frequency) is divided by n. In this case, relative frequency is evaluated for each interval, and the summation of all the heights of rectangles is equal to unity.

Frequency Density Histogram

A frequency density histogram is obtained taking the heights of the relative frequency histogram and dividing them by the interval size Δx. It constitutes the sample analogous of a PDF. In fact, the area of the ith rectangle corresponds to the relative frequency of the ith interval, and the summation of all the areas is equal to unity.

B.1.1.2. Plotting Positions

Plotting positions are the sample analogous of CDF values. First, the sample data are sorted in ascending order, and then the rank position i is evaluated for each value. In this case, i corresponds to the number of data which are less or equal to an assigned value. Dividing i by n, a cumulative frequency is obtained, which constitutes a first estimation of plotting positions $PP(i)$:

$$PP(i) = \frac{i}{n} \qquad i = 1, 2, \ldots, n \tag{B.5}$$

Using Eq. (B.5), a non-exceedance frequency equal to one is obtained for the maximum value, i.e. $i = n$. This means the event is certain to happen, which is not acceptable. In addition, the value of new future sample data could also exceed the maximum value. For this reason, a more general expression is adopted, which ensures that the cumulative frequency is always less than one:

$$PP(i) = \frac{i - a}{n + 1 - 2a} \qquad i = 1, 2, \ldots, n \tag{B.6}$$

where a is a constant. If $a = 0$, the Weibull plotting position is obtained:

$$PP(i) = \frac{i}{n+1} \qquad i = 1, 2, \dots, n \qquad \text{(B.7)}$$

If $a = 0.5$, the Hazen plotting position is obtained:

$$PP(i) = \frac{i - 0.5}{n} \qquad i = 1, 2, \dots, n \qquad \text{(B.8)}$$

If $a = 0.44$, the Gringorten plotting position is obtained:

$$PP(i) = \frac{i - 0.44}{n + 0.12} \qquad i = 1, 2, \dots, n \qquad \text{(B.9)}$$

Eqs. (B.7)-(B.9) produce similar results, especially when n number of data points is large.

B.1.2. Sample Moments

Like PDF and CDF, sample analogous of theoretical moments can be evaluated by the available data. In detail, it is possible to estimate sample versions of expected value, m_X :

$$m_X = \frac{1}{n} \sum_{i=1}^{n} x_i \qquad \text{(B.10)}$$

variance, s_X^2 :

$$s_X^2 = \frac{1}{n} \sum_{i=1}^{n} (x_i - m_X)^2 \qquad \text{(B.11)}$$

skewness, g_1 :

$$g_1 = \frac{(1/n)\sum_{i=1}^{n}(x_i - m_X)^3}{s_X^3}$$

(B.12)

and kurtosis, g_2 :

$$g_2 = \frac{(1/n)\sum_{i=1}^{n}(x_i - m_X)^4}{s_X^4} - 3$$

(B.13)

B.2. PARAMETER ESTIMATION

Once a theoretical distribution is hypothesized to be used for the available sample data, the parameter estimation can be carried out by considering several techniques.

In the following, two methods are described:

1.   Method of the Moments (Section B.2.1)
2.   Method of the Maximum Likelihood (Section B.2.2)

B.2.1. Method of the Moments

The method is based on the hypothesis of equality among sample moments and theoretical ones, which are functions of the parameters to be estimated. The number of equalities corresponds to the number of probability distribution parameters. Moreover, equalities are fixed starting from the lowest order moment.

In details, if a probability distribution is characterized by p parameters $(\theta_1, \theta_2, ..., \theta_p)$, Eqs. (A.51)-(A.52) of the theoretical moments (respectively, for discrete and continuous random variables) can be rewritten as:

$$\mu_X^{(R)} = \mu_X^{(R)}(\theta_1, \theta_2, ..., \theta_p) = \sum_{i=1}^{k} x_i^R f_X(x_i, \theta_1, \theta_2, ..., \theta_p)$$

(B.14)

$$\mu_X^{(R)} = \mu_X^{(R)}(\theta_1,\theta_2,...,\theta_p) = \int_{-\infty}^{+\infty} x^R f_X(x,\theta_1,\theta_2,...,\theta_p)dx \qquad (B.15)$$

where $R = 1,2,...,p$

Indicating the sample moments as:

$$M_R = \sum_{i=1}^{n} \frac{x_i^R}{n} \qquad (B.16)$$

the parameter estimation is carried out by solving the following system of equations:

$$\begin{cases} M_1 = \mu_X^{(1)}(\theta_1,\theta_2,...,\theta_p) \\ ... \\ M_p = \mu_X^{(p)}(\theta_1,\theta_2,...,\theta_p) \end{cases} \qquad (B.17)$$

B.2.2. Method of the Maximum Likelihood

Let $x_1,x_2,...,x_n$ be sample data, hypothesized to be derived from a probability distribution $f_X(x,\theta_1,\theta_2,...,\theta_p) = f_X(x,\underline{\theta})$. If $x_1,x_2,...,x_n$ are independent, then a probability measure of obtaining the data sample from $f_X(x,\underline{\theta})$ is provided by the likelihood function:

$$L(x_1,x_2,...,x_n,\underline{\theta}) = f_X(x_1,\underline{\theta})f_X(x_2,\underline{\theta})\cdots f_X(x_n,\underline{\theta}) = \prod_{i=1}^{n} f_X(x_i,\underline{\theta}) \quad (B.18)$$

The method of the maximum likelihood consists in estimation of $\hat{\underline{\theta}}$ parameter vector, which maximizes $L(x_1,x_2,...,x_n,\underline{\theta})$. It must be highlighted that $L(x_1,x_2,...,x_n,\underline{\theta})$ corresponds to the probability of obtaining $x_1,x_2,...,x_n$ for a discrete random variable, and it is proportional to this probability for a continuous random variable.

For the case of one parameter θ_1, the estimation is carried out by solving the following equation:

$$\frac{dL(x_1, x_2, ..., x_n, \theta_1)}{d\theta_1} = 0 \tag{B.19}$$

For p parameters, the following system of equations has to be solved:

$$\begin{cases} \dfrac{\partial L(x_1, x_2, ..., x_n, \theta_1, ..., \theta_p)}{\partial \theta_1} = 0 \\[2ex] \dfrac{\partial L(x_1, x_2, ..., x_n, \theta_1, ..., \theta_p)}{\partial \theta_2} = 0 \\[1ex] ... \\[1ex] \dfrac{\partial L(x_1, x_2, ..., x_n, \theta_1, ..., \theta_p)}{\partial \theta_p} = 0 \end{cases} \tag{B.20}$$

In practice, it is often more convenient to work with the logarithm of the likelihood function $\ln L(x_1, x_2, ..., x_n, \underline{\theta})$, called the log-likelihood function:

$$\ln L(x_1, x_2, ..., x_n, \underline{\theta}) = \sum_{i=1}^{n} \ln[f_X(x_i, \underline{\theta})] \tag{B.21}$$

This is due to the property that $L(x_1, x_2, ..., x_n, \underline{\theta})$ and $\ln L(x_1, x_2, ..., x_n, \underline{\theta})$ assume the maximum value in correspondence of the same vector $\hat{\underline{\theta}}$. Consequently, evaluation of partial derivatives for a summation is easier than for a product of functions.

B.3. MODEL VALIDATION

Model validation is aimed at testing if the hypothesis about the adopted theoretical probability distribution cannot be rejected or otherwise. In statistics, the term "not rejected" is preferred with respect to "accepted". This is because if a hypothesis is said to be accepted, then it seems that other plausible hypotheses cannot be used.

The validation can be carried out using a graphical approach or an analytical approach.

The graphical approach is easier to apply, and probabilistic papers are used. A probabilistic paper is defined for each theoretical probability distribution. It is constructed by applying a mathematical deformation of the $F_X(x)$ axis (and of the x axis if it is necessary, like the Weibull distribution, see Section B.3.1.1), so that a straight line can be obtained. The goal is to check the fitting of extreme values, i.e. the tails of the distribution. The details regarding the construction of a generic probabilistic paper are reported in Section B.3.1. As examples, the cases for Exponential, Weibull and Normal distributions are discussed in Section B.3.1.1.

Another way to carry out the model validation is the analytical approach, namely a statistical hypothesis test. A brief overview of this test is reported in Section B.3.2.

B.3.1. Construction of a Probabilistic Paper

A probabilistic paper can be constructed using the following steps:

1. Choose a mathematical deformation, which represents the theoretical distribution.
2. Superimpose sample data onto the deformed plot.

For the first step, a new variable Y is defined, characterized by a linear relationship with the random variable X:

$$Y = a + bX \tag{B.22}$$

in which a and b are functions of the estimated parameters related to the hypothesized theoretical distribution. It is possible to demonstrate that if a linear relationship like Eq.(B.22) exists then:

$$F_Y(y) = F_X(x) \tag{B.23}$$

Eq. (B.23) allows the construction of a probabilistic paper by considering two vertical axes. The axis on the right of the plot is for the Y values. The axis on the left of the plot is for the corresponding $F_Y(y)$ values. The $F_Y(y)$ axis therefore cannot be linear. With this probabilistic paper, the plotting of the theoretical distribution gives a straight line (Eq. B.22), and the $F_Y(y)$ values are the same as the $F_X(x)$ values (Eq. B.23). For the second step, the plotting positions of the sorted sample data are considered (Section B.1.1.2). As they are the sample analogous of $F_X(x)$, for the ith plotting position $PP(i)$, it is possible to define a sample value of Y (indicated as $y_S(i)$) using the inversion of Eq. (B.23):

$$y_S(i) = F_Y^{-1}(PP(i)) \tag{B.24}$$

To sum up, in a probabilistic plot, the theoretical distribution is represented by a straight line, while the sample data is represented by a set of points (Figure B.1). If the straight line fits well with the sample data, then the hypothesis about the theoretical distribution cannot be rejected.

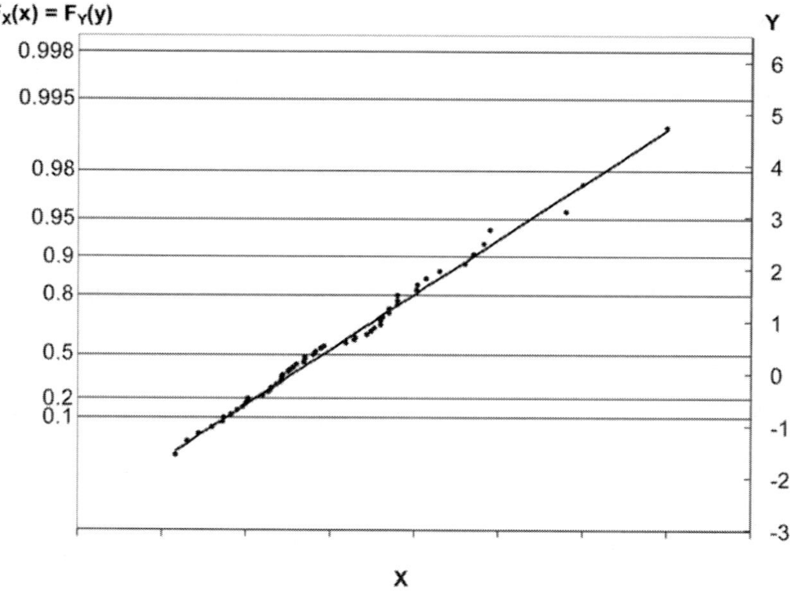

Figure B.1. Qualitative example of probabilistic paper and superimposing of sample data.

B.3.1.1. Examples of Probabilistic Papers

Exponential distribution

Starting from the CDF of an Exponential distribution:

$$F_X(x) = 1 - e^{-\lambda x} \tag{B.25}$$

the following expression for the variable Y is considered:

$$Y = \lambda X \tag{B.26}$$

resulting in:

$$F_Y(y) = 1 - e^{-y} \tag{B.27}$$

From Eq. (B.27)

$$y = -\ln\left(1 - F_Y(y)\right) \tag{B.28}$$

Table B.1 contains selected $F_Y(y)$ values and the corresponding y values according to Eq. (B.28).

Table B.1. Selected $F_Y(y)$ and y values for an Exponential probability paper

$F_Y(y)$	0.1	0.2	0.5	0.8	0.9	0.95	0.98	0.995	0.998
y	0.105	0.223	0.693	1.609	2.303	2.996	3.912	5.298	6.215

Weibull distribution

Starting from the CDF of a Weibull distribution:

$$F_X(x) = 1 - e^{-\alpha x^\beta} \tag{B.29}$$

a linear relationship is obtained by considering $\ln X$:

$$Y = \ln\left(\alpha X^{\beta}\right) = \ln \alpha + \beta \ln X \tag{B.30}$$

resulting in:

$$F_Y(y) = 1 - e^{-e^y} \tag{B.31}$$

From Eq. (B.31)

$$y = \ln\left(-\ln\left(1 - F_Y(y)\right)\right) \tag{B.32}$$

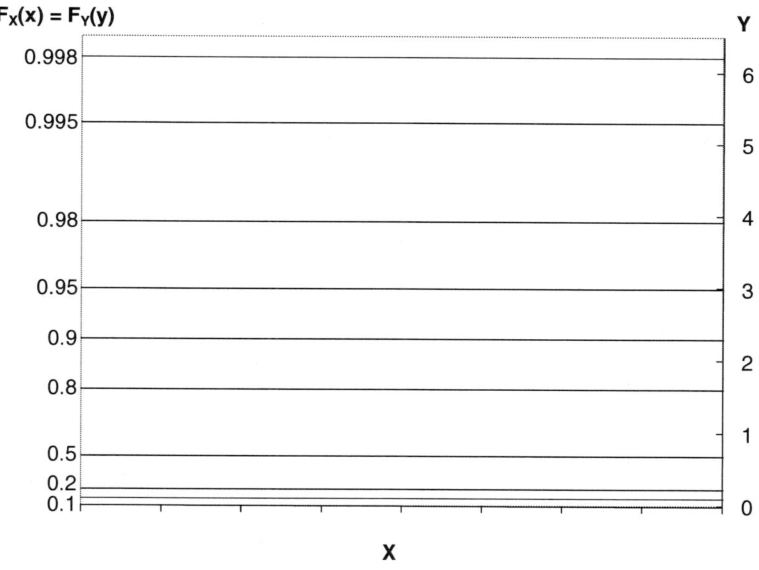

Figure B.2. Probabilistic paper for an Exponential distribution.

Table B.2 contains selected $F_Y(y)$ values and the corresponding y values according to Eq. (B.32).

Table B.2. Selected $F_Y(y)$ and y values for a Weibull probability paper

$F_Y(y)$	0.1	0.2	0.5	0.8	0.9	0.95	0.98	0.995	0.998
y	-2.250	-1.500	-0.367	0.476	0.834	1.097	1.364	1.667	1.827

Figure B.3. Probabilistic paper for a Weibull distribution.

Normal distribution

Starting from the CDF of a Normal distribution:

$$F_X(x) = \frac{1}{\sqrt{2\pi\sigma^2}} \int_{-\infty}^{x} e^{-\frac{(x-\mu)^2}{2\sigma^2}}$$

(B.33)

the following expression for the variable Y is considered:

$$Y = \frac{X-\mu}{\sigma}$$

(B.34)

resulting in:

$$F_Y(y) = \frac{1}{\sqrt{2\pi}} \int_{-\infty}^{y} e^{-\frac{y^2}{2}}$$

(B.35)

From Eq. (B.35):

$$y = Q^{-1}\big(F_Y(y)\big) \tag{B.36}$$

where Q^{-1} is the inverse of Q (Standard normal CDF), and it can be evaluated numerically using the numerical tools. Table B.3 contains selected $F_Y(y)$ values and the corresponding y values according to Eq. (B.36).

Table B.3. Selected $F_Y(y)$ and y values for Normal probability paper

$F_Y(y)$	0.1	0.2	0.5	0.8	0.9	0.95	0.98	0.995	0.998
y	-1.282	-0.842	0.000	0.842	1.282	1.645	2.053	2.576	2.878

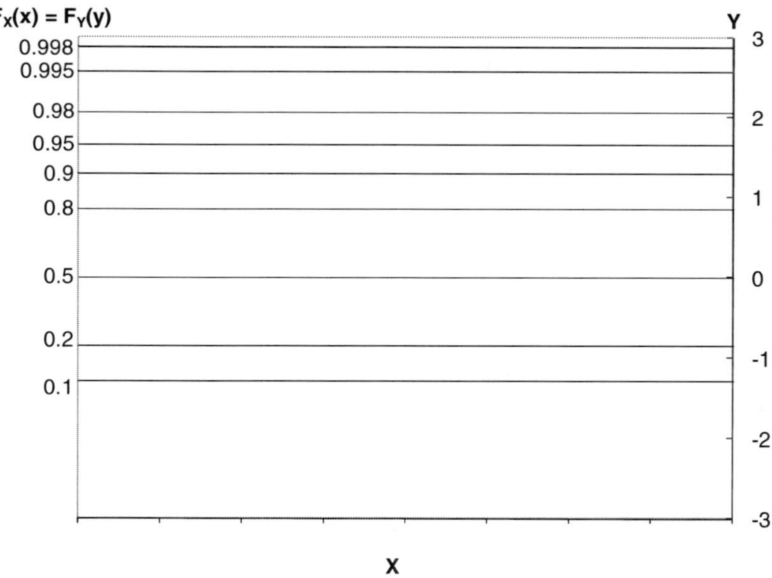

Figure B.4. Probabilistic paper for a Normal distribution.

B.3.2. Statistical Hypothesis Test

A null hypothesis, H_0, is formulated. It can be related to a whole set of sample data or to a single quantity of interest such as a parameter, a moment, etc. Examples of H_0 are:

1. an assigned theoretical probability distribution gives good reproduction of the entire set of sample data;
2. the single quantity of interest assumes a particular value or can be reproduced by an assigned model.

Then, on the basis of the available data, an analytical test can be carried out in order to decide whether H_0 cannot be rejected or otherwise. Table B.4 summarizes the four possible outcomes of the statistical hypothesis test.

Table B.4. Four possible outcomes of the statistical hypothesis test

	H_0 is true	H_0 is false
H_0 is not rejected	Correct decision	Wrong decision Type II error
H_0 is rejected	Wrong decision Type I error	Correct decision

A type I error occurs when H_0 is true, but it is rejected. A type II error occurs when H_0 is false, but it is erroneously not rejected. In this context, a *significance level* is defined. It is equal to the probability α associated with the rejection of a true hypothesis (type I error). Moreover, β is defined as the probability associated to the non-rejection of a false hypothesis (type II error).

In order to have a reliable statistical hypothesis test, the probabilities associated with the wrong decisions have to be minimized. It is clear that a decrease of α induces an increase of β and vice versa. In general, a good compromise is setting $\alpha = 0.05$.

On the basis of the value of α, a non-rejection region (also named confidence interval if the test regards a single quantity) is defined. It indicates the range in which the test result should be comprised for not rejecting the hypothesis H_0, with a probability $(1-\alpha)$ of not making a type I error. Moreover, it is possible to define a one-tailed test and a two-tailed test. For the one-tailed test, the non-rejection region excludes only the results too small or too large. For the two-tailed test, both too small and too large results are excluded.

If H_0 is related to the entire set of sample data then technical literature proposes several tests, from which we can recommend the χ^2 test and the Kolmogorov-Smirnov test (Kottegoda and Rosso 1997).

If H_0 regards a single quantity, examples of confidence intervals are reported in this book. In detail:

1. In Figure 3.3 of Chapter 3, the 95% percentiles of PRAISE model represent the upper boundary of the one-tailed confidence intervals, with $\alpha = 0.05$. The hypothesis H_0 is "the future observed values of rainfall heights are reproduced by PRAISE model". A similar example is in Figure 3.8 of Chapter 3, related to PRAISEST model.

2. Figure 3.7 of Chapter 3 and Figures 5.3-5.4 of Chapter 5 show two-tailed confidence intervals related to $\alpha = 0.05$. In this case, the boundaries of the interval are the 2.5% and 97.5% percentiles. The difference between these probability values is equal to 95%, which represents the probability $(1 - \alpha)$ of not making a type I error. In these cases the hypothesis H_0 is "the quantity of interest (mean, standard deviation, etc.) is reproduced by the adopted model".

B.4. REGRESSION

By taking the sample data, regression is a technique that attempts to model the relationship between two or more random variables. If the relationship is related to only two variables *(X, Y)*, then regression is said to be 'simple'. On the other hand, if the relationship regards more than two variables, then regression is said to be 'multiple'. Depending on the structure of mathematical relationships, regression is said to be 'linear' or 'non-linear'. First, the case of simple linear regression is analyzed. Then, extensions to more general cases are discussed.

B.4.1. Simple Linear Regression

In the simple regression, X and Y represent the independent and the dependent random variable, respectively. The general relationship between them is:

$$Y = f(X) + e \tag{B.37}$$

where e is the error (i.e. the difference between the observations and the theoretical values of Y).

In the case of a simple linear regression, Eq. (B.37) is expressed as:

$$Y = a_0 + a_1 X + e \tag{B.38}$$

Starting from n pairs of observations (x_i, y_i), $i = 1, 2, ..., n$, Eq. (B.38) can be rewritten as (Figure B.5):

$$y_i = a_0 + a_1 x_i + e_i \tag{B.39}$$

The coefficients a_0 and a_1 can be estimated by using the ordinary least squares method, i.e. by minimizing the following function S:

$$S = \sum_{i=1}^{n} e_i^2 = \sum_{i=1}^{n} \left(y_i - y_i^*\right)^2 = \sum_{i=1}^{n} \left(y_i - a_0 - a_1 x_i\right)^2 \tag{B.40}$$

in which $y_i^* = a_0 + a_1 x_i$ is the ith theoretical value for the random variable Y. In detail, partial derivatives of S with respect to a_0 and a_1 are evaluated and then fixed equal to zero:

$$\begin{cases} \dfrac{\partial S}{\partial a_0} = -2 \sum_{i=1}^{N} \left(y_i - a_0 - a_1 x_i\right) = 0 \\[2em] \dfrac{\partial S}{\partial a_1} = -2 \sum_{i=1}^{N} \left(y_i - a_0 - a_1 x_i\right) x_i = 0 \end{cases} \tag{B.41}$$

From Eq. (B.41), the following linear system of equations is obtained:

$$\begin{cases} a_0 n + a_1 \sum_{i=1}^{n} x_i = \sum_{i=1}^{n} y_i \\[2em] a_0 \sum_{i=1}^{n} x_i + a_1 \sum_{i=1}^{n} x_i^2 = \sum_{i=1}^{n} x_i y_i \end{cases} \tag{B.42}$$

The expressions for a_0 and a_1 are:

$$a_0 = \frac{\left(\sum_{i=1}^{n} y_i\right)\left(\sum_{i=1}^{n} x_i^2\right) - \left(\sum_{i=1}^{n} x_i\right)\left(\sum_{i=1}^{n} x_i y_i\right)}{n\left(\sum_{i=1}^{n} x_i^2\right) - \left(\sum_{i=1}^{n} x_i\right)^2} \qquad (B.43)$$

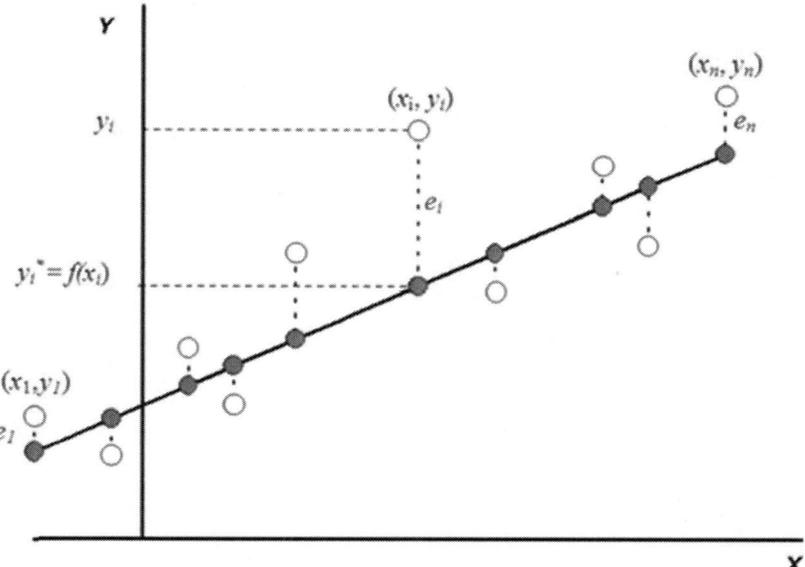

Figure B.5. Example of a linear regression into a scatter plot.

$$a_1 = \frac{n\left(\sum_{i=1}^{n} x_i y_i\right) - \left(\sum_{i=1}^{n} x_i\right)\left(\sum_{i=1}^{n} y_i\right)}{n\left(\sum_{i=1}^{n} x_i^2\right) - \left(\sum_{i=1}^{n} x_i\right)^2} \qquad (B.44)$$

For the available set of n observed pairs (x_i, y_i), the simple linear regression is characterized by the following properties:

1. The theoretical straight line $Y^* = a_0 + a_1 X$ is unique and comprises the pair (m_X, m_Y), i.e. the sample expected values.

2. The coefficients a_0, a_1 are such that the property $\sum_{i=1}^{n} e_i = 0$ is satisfied.

This is the first equation of the system in (B.41).

Moreover, Eqs. (B.43)- (B.44) can be rewritten as:

$$a_1 = \frac{\dfrac{1}{n}\sum_{i=1}^{n}(x_i - m_X)\cdot(y_i - m_Y)}{\dfrac{1}{n}\sum_{i=1}^{n}(x_i - m_X)^2} = \frac{cov[X,Y]}{s_X^2} \tag{B.45}$$

$$a_0 = m_Y - a_1 m_X \tag{B.46}$$

The numerator of Eq. (B.45) corresponds to the sample evaluation of the theoretical covariance $Cov[X,Y]$ (see Section A.3.6). Consequently, it is possible to define the sample coefficient of correlation $r_{X,Y}$ as:

$$r_{X,Y} = \frac{cov[X,Y]}{s_X s_Y} = \frac{\dfrac{1}{n}\sum_{i=1}^{n}(x_i - m_X)(y_i - m_Y)}{\sqrt{\dfrac{1}{n}\sum_{i=1}^{n}(x_i - m_X)^2 \dfrac{1}{n}\sum_{i=1}^{n}(y_i - m_Y)^2}} \tag{B.47}$$

B.4.2. Examples of Simple Non-Linear Regression

As examples of simple non-linear regression, first the parabolic case is described, i.e.:

$$Y = a_0 + a_1 X + a_2 X^2 + e \tag{B.48}$$

Starting from n pairs of observations (x_i, y_i), $i = 1, 2, ..., n$, Eq. (B.48) can be rewritten as:

$$y_i = a_0 + a_1 x_i + a_2 x_i^2 + e_i \tag{B.49}$$

The coefficients a_0, a_1 and a_2 can be estimated using the ordinary least squares method, i.e. by minimizing the following function S:

$$S = \sum_{i=1}^{n} e_i^2 = \sum_{i=1}^{n} \left(y_i - y_i^* \right)^2 = \sum_{i=1}^{n} \left(y_i - a_0 - a_1 x_i - a_2 x_i^2 \right)^2 \qquad (B.50)$$

in which $y_i^* = a_0 + a_1 x_i + a_2 x_i^2$ is the ith theoretical value for the random variable Y. Partial derivatives of S with respect to a_0, a_1 and a_2 are evaluated and fixed equal to zero:

$$\begin{cases} \dfrac{\partial S}{\partial a_0} = -2 \sum_{i=1}^{N} \left(y_i - a_0 - a_1 x_i - a_2 x_i^2 \right) = 0 \\[2em] \dfrac{\partial S}{\partial a_1} = -2 \sum_{i=1}^{N} \left(y_i - a_0 - a_1 x_i - a_2 x_i^2 \right) x_i = 0 \\[2em] \dfrac{\partial S}{\partial a_2} = -2 \sum_{i=1}^{N} \left(y_i - a_0 - a_1 x_i - a_2 x_i^2 \right) x_i^2 = 0 \end{cases} \qquad (B.51)$$

From (B.51) the following linear system of equations is obtained:

$$\begin{cases} a_0 n + a_1 \sum_{i=1}^{n} x_i + a_2 \sum_{i=1}^{n} x_i^2 = \sum_{i=1}^{n} y_i \\[2em] a_0 \sum_{i=1}^{n} x_i + a_1 \sum_{i=1}^{n} x_i^2 + a_2 \sum_{i=1}^{n} x_i^3 = \sum_{i=1}^{n} x_i y_i \\[2em] a_0 \sum_{i=1}^{n} x_i^2 + a_1 \sum_{i=1}^{n} x_i^3 + a_2 \sum_{i=1}^{n} x_i^4 = \sum_{i=1}^{n} x_i^2 y_i \end{cases} \qquad (B.52)$$

which can be organized in the matrix form:

$$\underline{\underline{M}}\, \underline{A} = \underline{B} \qquad (B.53)$$

with

$$\underline{\underline{M}} = \begin{bmatrix} n & \sum_{i=1}^{n} x_i & \sum_{i=1}^{n} x_i^2 \\ \sum_{i=1}^{n} x_i & \sum_{i=1}^{n} x_i^2 & \sum_{i=1}^{n} x_i^3 \\ \sum_{i=1}^{n} x_i^2 & \sum_{i=1}^{n} x_i^3 & \sum_{i=1}^{n} x_i^4 \end{bmatrix} \quad \underline{A} = \begin{bmatrix} a_0 \\ a_1 \\ a_2 \end{bmatrix} \quad \underline{B} = \begin{bmatrix} \sum_{i=1}^{n} y_i \\ \sum_{i=1}^{n} x_i y_i \\ \sum_{i=1}^{n} x_i^2 y_i \end{bmatrix} \quad (B.54)$$

It can be noted that $\underline{\underline{M}}$ is a symmetric matrix. Extensions to cubic or polynomial regression of higher order are straightforward. As a second example of simple non-linear regression, the power law case can be mentioned:

$$Y = a_0 X^{a_1} + e \qquad (B.55)$$

By considering the theoretical value $Y^* = a_0 X^{a_1}$, the corresponding logarithmic transformation is:

$$\ln Y^* = \ln\left(a_0 X^{a_1}\right) = \ln a_0 + a_1 \ln X \Rightarrow LY = A_0 + a_1 LX \qquad (B.56)$$

with $A_0 = \ln a_0$, $LY = \ln Y^*$ and $LX = \ln X$. Thus, starting from n pairs of observations (x_i, y_i), $i = 1, 2, ..., n$, the simple linear regression can be adopted for the transformed observations $(\ln x_i, \ln y_i)$, in order to estimate A_0 and a_1; consequently $a_0 = e^{A_0}$.

B.4.3. Examples of Multiple Linear Regression

As an example of multiple linear regression, the trivariate case is described, i.e.:

$$Z = a_0 + a_1 X + a_2 Y + e \qquad (B.57)$$

Starting from n sets of observations (x_i, y_i, z_i), $i = 1, 2, ..., n$, Eq. (B.57) can be rewritten as:

$$z_i = a_0 + a_1 x_i + a_2 y_i + e_i \tag{B.58}$$

The coefficients a_0, a_1 and a_2 can be estimated using the ordinary least squares method, i.e. by minimizing the following function:

$$S = \sum_{i=1}^{n} e_i^2 = \sum_{i=1}^{n} \left(z_i - z_i^*\right)^2 = \sum_{i=1}^{n} \left(z_i - a_0 - a_1 x_i - a_2 y_i\right)^2 \tag{B.59}$$

in which $z_i^* = a_0 + a_1 x_i + a_2 z_i$ is the ith theoretical value for the random variable Z. Partial derivatives of S with respect to a_0, a_1 and a_2 are evaluated and fixed equal to zero:

$$\begin{cases} \dfrac{\partial S}{\partial a_0} = -2 \sum_{i=1}^{N} \left(z_i - a_0 - a_1 x_i - a_2 y_i\right) = 0 \\[2ex] \dfrac{\partial S}{\partial a_1} = -2 \sum_{i=1}^{N} \left(z_i - a_0 - a_1 x_i - a_2 y_i\right) x_i = 0 \\[2ex] \dfrac{\partial S}{\partial a_2} = -2 \sum_{i=1}^{N} \left(z_i - a_0 - a_1 x_i - a_2 y_i\right) y_i = 0 \end{cases} \tag{B.60}$$

From Eq. (B.60), the following linear system of equations is obtained:

$$\begin{cases} a_0 n + a_1 \sum_{i=1}^{n} x_i + a_2 \sum_{i=1}^{n} y_i = \sum_{i=1}^{n} z_i \\[2ex] a_0 \sum_{i=1}^{n} x_i + a_1 \sum_{i=1}^{n} x_i^2 + a_2 \sum_{i=1}^{n} y_i x_i = \sum_{i=1}^{n} z_i x_i \\[2ex] a_0 \sum_{i=1}^{n} y_i + a_1 \sum_{i=1}^{n} y_i x_i + a_2 \sum_{i=1}^{n} y_i^2 = \sum_{i=1}^{n} z_i y_i \end{cases} \tag{B.61}$$

which can be organized in the matrix form:

$$\underline{\underline{M}}\,\underline{A} = \underline{B} \tag{B.62}$$

with

$$\underline{\underline{M}} = \begin{bmatrix} n & \sum_{i=1}^{n} x_i & \sum_{i=1}^{n} y_i \\ \sum_{i=1}^{n} x_i & \sum_{i=1}^{n} x_i^2 & \sum_{i=1}^{n} y_i x_i \\ \sum_{i=1}^{n} y_i & \sum_{i=1}^{n} y_i x_i & \sum_{i=1}^{n} y_i^2 \end{bmatrix} \quad \underline{A} = \begin{bmatrix} a_0 \\ a_1 \\ a_2 \end{bmatrix} \quad \underline{B} = \begin{bmatrix} \sum_{i=1}^{n} z_i \\ \sum_{i=1}^{n} z_i x_i \\ \sum_{i=1}^{n} z_i y_i \end{bmatrix} \tag{B.63}$$

It can be noted that $\underline{\underline{M}}$ is a symmetric matrix. Extensions to multivariate linear regression cases are straightforward.

B.4.4. Examples of Multiple Non-Linear Regression

As an example of multiple not-linear regressions, the power law case can be mentioned:

$$Z = a_0 X^{a_1} Y^{a_2} + e \tag{B.64}$$

By considering the theoretical value $Z^* = a_0 X^{a_1} Y^{a_2}$, the corresponding logarithmic transformation is:

$$\ln Z^* = \ln\left(a_0 X^{a_1} Y^{a_2}\right) = \ln a_0 + a_1 \ln X + a_2 \ln Y \Rightarrow LZ = A_0 + a_1 LX + a_1 LY \tag{B.65}$$

with $A_0 = \ln a_0$, $LZ = \ln Z^*$, $LY = \ln Y$ and $LX = \ln X$.

Thus, starting from n sets of observations (x_i, y_i, z_i), $i = 1, 2, ..., n$, the multiple linear regression can be adopted for the transformed observations $(\ln x_i, \ln y_i, \ln z_i)$ in order to estimate A_0, a_1 and a_2; consequently $a_0 = e^{A_0}$.

B.5. Validation Tests for Meta-Gaussian Model

The Meta-Gaussian model is described in Section 3.2.4 of Chapter 3. Herr and Krzysztofowicz (2005) showed that the tests for validating a Meta-Gaussian model are related to:

1. Linearity of the regression between the variables Y on X;
2. Homoscedasticity of the residual $\Theta = Y - \rho \cdot X$;
3. The distribution of Θ has to be normal with mean equal to zero and the variance equal to $1 - \rho^2$.

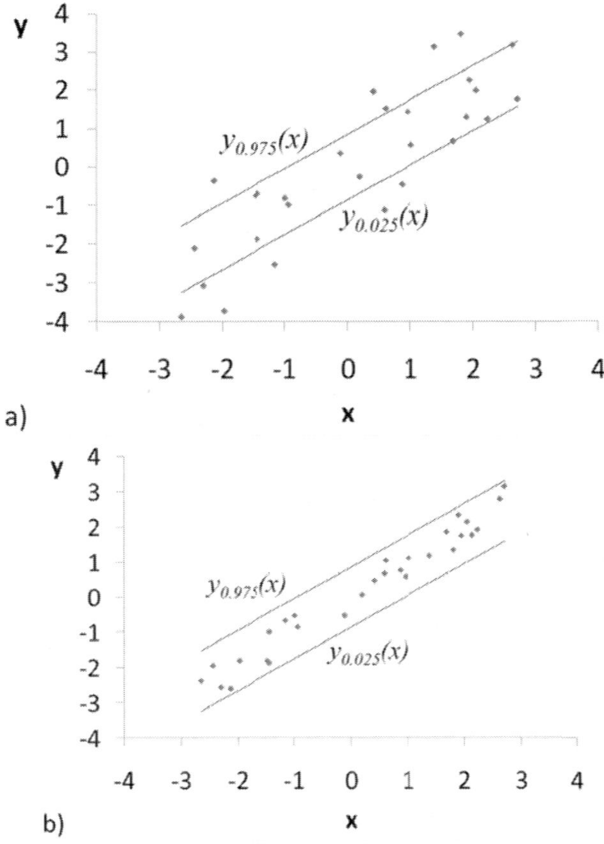

Figure B.6. Linearity of regression: qualitative examples of a) rejection and b) non-rejection of the hypothesis.

The linearity of the regression can be tested in a graphical way. In the scatter plot of the variables Y and X the sample points, the theoretical straight line of the regression and its correspondent non-rejection region are represented. In detail, the non-rejection region is a two-tailed region, for which the lower and upper boundaries can be evaluated as follows. The p percentile of Y, conditioned on $X = x$ is

$$y_p(x) = \rho x + \sqrt{1-\rho^2}\, Q^{-1}(p) \tag{B.65}$$

Figure B.6 shows qualitative examples of rejection and non-rejection of the hypothesis.

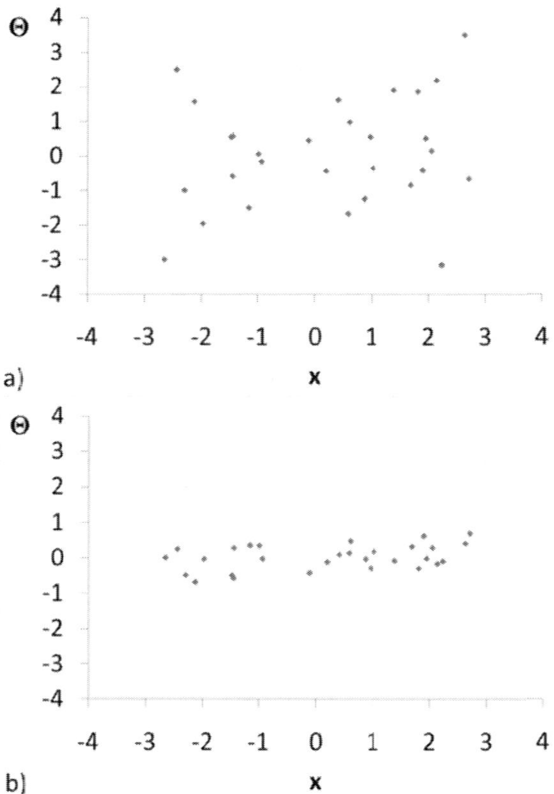

Figure B.7. Homoscedasticity of the residual Θ : qualitative examples of a) rejection and b) non-rejection of the hypothesis.

Homoscedasticity means that the variance of the residuals has to be independent on X. Also this test can be carried out in a graphical way. Figure B.7 shows qualitative examples of rejection and non-rejection of the hypothesis.

Finally, the hypothesis of a normal distribution for Θ with assigned values of the parameters can be validated using a normal probabilistic paper (see Section 3.1.1.1 for further details about the construction of a probabilistic paper).

REFERENCES

Abraham, B. (1983). "The exact likelihood function for space-time model." *Metrika*, Vol. 30, pp. 239–243.

Abramowitz, M. and Stegun, I. A. (1970). *"Handbook of mathematical functions"*, Dover, New York, NY.

Anthes, R.A., and Warner, T.J. (1978). "Development of hydrodynamic models suitable for air pollution and other mesometeorological studies." *Monthly Weather Review*, Vol. 108, pp. 1045–1078.

Antolik, M.S. (2000). "An overview of the National Weather Service's centralized statistical quantitative precipitation forecasts." *Journal of Hydrology*, Vol. 239, No. 1–4, pp. 306–337.

Aronian, L. A. (1980). "Time series in m dimensions: Definitions, problems and prospects." *Communications in Statistics, Simulation and Computation*, Vol. B9, pp. 453–465.

Austin, P. M., and Houze, Jr., R.A. (1972). "Analysis of the structure of the precipitation patterns in New England." *Journal of Applied Meteorology*, Vol. 11, pp. 926–934.

Baum, R. L., Godt, J. W., Harp, E. L., McKenna, J. P. and McMullen, S. R. (2005). "Early warning of landslides for rail traffic between Seattle and Everett, Washington, USA." Proceedings of *Conference on Landslide Risk Management* (Vancouver), pp. 731–740.

Barbe, P., Genest, C., Ghoudi, K., and Rémillard, B. (1996). "On Kendall's process." *Journal of Multivariate Analysis*, Vol. 58, pp. 197–229.

Bergthorsson, P., and Doos, B. (1955). "Numerical weather map analysis". Tellus, Vol. 7, No. 3, pp. 329-340.

Beven, K. J., and Bynley, A. M. (1992). "The Future of Distributed Models: Model Calibration and Predictive Uncertainty." *Hydrological Processes*, Vol. 6, pp. 279–298.

Box, G. E. P., and Cox, D. R. (1964). "An analysis of transformations." *Journal of Royal Statistical Society, Series B*, Vol. 26, pp. 211–252.

Box, G. E. P., and Jenkins, G. M. (1976). *"Time series analysis. Forecasting and Control."* Holden Day, San Francisco, CA.

Bras, R. L., and Rodriguez-Iturbe, I. (1994). *"Random Functions and Hydrology."* Dover, New York, NY.

Brockwell, P. J., and Davis R. A. (1987). *"Time Series: Theory and Methods."* Springer, New York, NY.

Burlando, P., and Rosso, R. (1993). "Stochastic models of temporal rainfall: reproducibility, estimation and prediction of extreme events." In *Stochastic Hydrology in its Use in Water Resources Systems Simulation and Optimization* (edited by Salas, J.D., R. Harboe, e J. Marco-Segura), Proceedings of NATO-ASI Workshop, Peniscola, Spain, September 18-29, 1989, Kluwer, pp. 137–173.

Burlando, P., Rosso, R., Cadavid, L. G., and Salas, J. D. (1993). "Forecasting of short-term rainfall using ARMA models." *Journal of Hydrology*, Vol. 144, pp. 193–211.

Burlando, P., Montanari, A. and Ranzi, R. (1996). "Forecasting of storm rainfall by combined use of radar, rain gages and linear models." Atmospheric Research, Vol. 42, pp. 199–216.

Cadavid, L., Salas, J. D., and Boes, D. C. (1992). "Disaggregation of precipitation records." *Water Resources Papers,* 106, Eng. Res. Center, Colorado State University, Fort Collins, CO.

Camacho, F., McLeod A. I. , and Hipel, K. W. (1985). "Developments in Multivariate ARMA Modeling in Hydrology." *International Symposium on Multivariate Analysis of Hydrologic Processes*, Colorado State University, Fort Collins, CO.

Capparelli, G., and Versace, P. (2011). "FLaIR and SUSHI": two mathematical models for early warning of landslides induced by rainfall." *Landslides*, Vol. 8, pp. 67–79.

Cliff, A. D., Hagget P., Ord, J. K., Bassett, K. A., and Davies R.B. (1975)*"Elements of Spatial Structure: A Quantitative Approach."* Cambridge University Press, UK.

Cole, K., and Davis, G. M. (2002). "Landslide warning and emergency planning systems in West Dorset, England." In: *Instability, Planning and*

Management (ed. by McInnes, R. G. and Jakeways, J.), Thomas Telford, London, UK, pp. 463–470.

Cox, D. R., and Isham, V. (1980). *"Point Processes."* Champan and Hall, London, UK.

Cressie, N. A. C. (1993). *Statistics for spatial data.* John Wiley and Sons, New York, NY.

Cressman, G. (1959) "An operational analysis system." *Monthly Weather Review*, Vol. 87, pp. 367-74.

Damrath, U., Doms, G., Fruhwald, D., Heise, E., Richter, B., and Steppeler, J. (2000). "Operational quantitative precipitation forecasting at the German Weather Service." *Journal of Hydrology*, Vol. 239, No. 1–4, pp. 260–285.

De Luca, D. L. (2006). *"Metodi di previsione dei campi di pioggia."* PhD Thesis (in italian), University of Calabria (Italy).

De Luca, D. L., and Versace, P. (2010). "Stochastic models for rainfall nowcasting." *Proceedings of STAHY 2010 International Workshop on Advances in Statistical Hydrology*, Taormina (Italy), 23-25 May 2010 (published on line: http://www.risorseidriche.dica.unict.it/Sito_STAHY 2010_web/proceedings.htm).

De Luca, D. L., Biondi, D., Capparelli, G., Galasso, L., and Versace, P. (2010). "Mathematical models for early warning systems." In *Global Change Facing Risks and Threats to Water Resources, Proceedings of the sixth World FRIEND Conference*, Fez (Morocco), 25-29 October 2010, pp. 485-495, IAHS Publications 340, Wallingford, UK.

De Michele, C., Salvadori, G., Passoni G., and Vezzosi, R. (2007). "A multivariate model of sea storms using copulas." *Coastal Engineering*, Vol. 54, pp. 734–751.

d'Orsi, R. N., D'Avila, C., Ortigao, J. A. R., Moraes, L., and Santos, M. D. (1997). "Rio-The Rio de Janeiro landslide watch." *Proceedings of 2nd PSL Pan-Am Symposium on Landslides Rio de Janeiro*, Vol. 1, pp. 21–30.

Duan, Q., Sorooshian, S., and Gupta, V. K. (1992). "Effective and efficient global optimization for conceptual rainfall-runoff models." *Water Resources Research*, Vol. 28, No. 4, pp. 1015-1031.

Entekhabi, D., Rodriguez-Iturbe, I., and Eagleson, P. S. (1989). "Probabilistic representation of the temporal rainfall process by a Modified Neyman-Scott Rectangular Pulse Model: parameter estimation validation." *Water Resources Research*, Vol. 25, No. 2, pp. 295–302.

French, M. N., Krajewski, W. F., and Cuykendal, R. R. (1992). "Rainfall forecasting in space and time using a neural network." *Journal of Hydrology*, Vol. 137, pp. 1–37.

Gandin, L. (1963). *"Objective analysis of meteorological fields (Leningrad:Gridromet)."* English translation (Jerusalem: Israel Program for Scientific Translation)

Genest, C., Ghoudi, K., and Rivest, L. P. (1995). "A semiparametric estimation procedure of dependence parameters in multivariate families of distributions." *Biometrika*, Vol. 82, No. 3, pp. 543–552.

Gilchrist, B., and Cressman, G. (1954). "An experiment in objective analysis." *Tellus*, Vol. 6, pp. 309-318.

Glade, T., Crozier, M., and Smith, P. (2000). "Applying probability determination to refine landslide triggering rainfall thresholds using an empirical Antecedent Daily Rainfall Model." *Pure and Applied Geophysics*, Vol. 157, No. 6–8, pp. 1059–1079.

Golding, B. W. (2000). "Quantitative precipitation forecasting in the UK." *Journal of Hydrology*, Vol. 239, No. 1–4, pp. 286–305.

Govindaraju, R. S. (2000a). "Artificial neural network in hydrology, I: Preliminary concepts." *Journal of Hydrology*, Vol. 5, No. 2, pp. 115–123.

Govindaraju, R. S. (2000b). "Artificial neural network in hydrology, II: Hydrological applications." *Journal of Hydrology*, Vol. 5, No. 2, pp. 124–137.

Grimaldi, S., and Serinaldi, F. (2006). "Design hyetograph analysis with 3-copula function." *Hydrological Sciences Journal*, Vol. 51, No. 2, pp. 223-238.

Herr, H. D., and Krzysztofowicz, R. (2005). "Generic probability distribution of rainfall in space: the bivariate model." *Journal of Hydrology*, Vol. 306, pp. 234–263.

Hipel, K. W., McLeod, A. I., and Lennox, W. C. (1977). "Advances in Box-Jenkins modelling 1: model construction." *Water Resources Research*, Vol. 13, No. 3, pp. 567–575.

Hsu, K., Gupta, H. V., Sorooshian, S., and Gao, X. (1996). "An artificial neural network for rainfall estimation from satellite infrared imaginary. Applications of remote sensing in hydrology." In *Proceedings of 3rd international workshop, NHRI symposium No. 17,* NASA, Greenbelt, Md.

Hsu, K., Gao, X., Sorooshian, S., and Gupta, H. (1997). "Precipitation estimation from remotely sensed information using artificial neural networks." *Journal of Applied Meteorology*, Vol. 36, No. 9, pp. 1176–1190.

Huard, D., and Mailhot, A. (2008). "Calibration of hydrological model GR2M using Bayesian uncertainty analysis." *Water Resources Research*, Vol. 44, W02424, doi: 10.1029/2007WR005949.

Johnson, E. R., and Bras, R. L. (1980). "Multivariate short-term rainfall prediction." *Water Resources Research*, Vol. 16, No. 7, pp. 173–185.

Joe, H. (1997). *"Multivariate models and dependence concepts."* Chapman and Hall, New York, NY.

Kayano, K., and Shimizu, K. (1994). "Optimal threshold for a mixture of lognormal distributions as the continuous part of a mixed distribution." *Journal of Applied Meteorology*, Vol. 33, pp. 1543–1550.

Kavetski, D., Kuczera, G., and Franks, S. W. (2006a). "Bayesian analysis of input uncertainty in hydrological modeling: 1. Theory." *Water Resources Research*, Vol. 42, W03407, doi:10.1029/2005WR004368.

Kavetski, D., Kuczera, G., and Franks, S. W. (2006b). "Bayesian analysis of input uncertainty in hydrological modeling: 2. Application." *Water Resources Research*, Vol. 42, W03408, doi:10.1029/2005WR004376.

Kedem, B., Chiu, L. S., and Karni, Z. (1990). "An analysis of the threshold method for measuring area-average rainfall." *Journal of Applied Meteorology*, Vol. 29, pp. 3–20.

Keefer, D. K., Wilson, R. C., Mark, R. K., Brabb, E. E., Brown III, W. M., Ellen, S. D., Harp, E. L., Wieczoreck, G. F., Alger, C. S., and Zatkin, R. S. (1987). "Real-time landslide warning during heavy rainfall." *Science*, Vol. 238, pp. 921–926.

Kelly, K. S., and Krzysztofowicz, R. (1997). "A bivariate meta-Gaussian density for use in hydrology." *Stochastic Hydrology and Hydraulics*, Vol. 11, pp. 17–31.

Kimberling, C. A. (1974). "Probabilistic interpretation of complete monotonicity." *Aequationes Mathematicæ*, Vol. 10, pp. 152–164.

Kingston, G. B., Lambert, M. F., and Maier, H. R. (2003). "Development of Stochastic Artificial Neural Networks for Hydrological Prediction." *Proceedings of International Congress on Modelling and Simulation MODSIM 2003 (Modelling and Simulation Society of Australia and New Zeeland)*, 14 – 17 July, pp. 837–842.

Kleiber, C, and Kotz, S. (2003). *"Statistical Size Distributions in Economics and Actuarial Sciences."* John Wiley and Sons, New York, NY.

Kottegoda, N. T., and Rosso R. (1997). *"Probability, statistics and reliability for civil and environmental engineers."* McGraw-Hill, New York, USA.

Kotz, S., Balakrishanan, N., and Johnson, N.L. (2000). *"Continuous Multivariate Distributions – Models And Applications."* John Wiley and Sons, New York, NY.

Kroese, D. P., Taimre, T., and Botev, Z. I. (2011). *"Handbook of Monte Carlo Methods"*. Wiley Series in Probability and Statistics, John Wiley and Sons, New York, NY.

Krzysztofowicz, R. (1999). "Bayesian theory of probabilistic forecasting via deterministic hydrologic model." *Water Resources Research*, Vol. 35, No. 9, pp. 2739–2750.

Krzysztofowicz, R. (2001). "The case for probabilistic forecasting in hydrology". *Journal of Hydrology*, Vol. 249, pp. 2–9.

Krzysztofowicz, R. (2002). "Bayesian system for probabilistic river stage forecasting." *Journal of Hydrology*, Vol. 268, pp. 16–40.

Krzysztofowicz, R., and Herr, H. (2001). "Hydrologic uncertainty processor for probabilistic river stage forecasting: precipitation-dependent model." *Journal of Hydrology*, Vol. 249, pp. 69–85.

Krzysztofowicz, R., Drzal, W. J., Drake, T. R., Weyman, J. C., and Giordano, L. A. (1993). "Probabilistic quantitative precipitation forecasts for river basin." *Weather and Forecasting*, Vol. 8, No. 4, pp. 424–439.

Kuczera, G., Kavetski, D., Franks, S., and Thyer, M. (2006). "Towards a Bayesian total error analysis of conceptual rainfall–runoff models: Characterising model error using storm-dependent parameters." *Journal of Hydrology*, Vol. 331, pp. 161–177.

Kuhn, G., Khan S., Ganguly, A. R., and Branstetter, M. L. (2007). "Geospatial-temporal dependence among weekly precipitation extremes with applications to observations and climate model simulations in South America." *Advances in Water Resources*, Vol. 30, pp. 2401–2423.

Kuligowski, R. J., and Barros, A. P. (1998a). "Localized precipitation forecasts from a numerical weather prediction model using artificial neural networks." *Weather and Forecasting*, Vol. 13, pp. 1195–1205.

Kuligowski, R. J., and Barros, A. P. (1998b). "Experiments in short-term precipitation forecasting using artificial neural networks". *Monthly Weather Review*, Vol. 126, No. 2, pp. 470–482.

Lin, G. F., and Chen, L. H. (2005). "Application of an artificial neural network to typhoon rainfall forecasting." *Hydrological Processes*, Vol. 19, pp. 1825–1837.

Liu, Y., and Gupta, H. V. (2007). "Uncertainty in hydrologic modeling: Toward an integrated data assimilation framework." *Water Resources Research*, Vol. 43, W07401, doi: 10.1029/2006WR005756.

Lorenz, E. N. (1963). "Deterministic nonperiodic flow". Journal of Atmospheric Sciences, Vol. 20, No. 2, pp. 130–141.

Luk, K. G., Ball, J. E., and Sharma, A. (2000). "Study of optimal lag and statistical inputs to artificial neural network for rainfall forecasting." *Journal of Hydrology*, Vol. 227, pp. 56–65.

Luk, K. G., Ball, J. E., and Sharma, A. (2001). "An application of artificial neural network for rainfall forecasting." *Mathematical and Computer Modeling*, Vol. 33, pp. 683–693.

Maier, H., and Dandy, G. (2000). "Neural networks for the predictions and forecasting of water resources variables: review of modelling issues and applications." *Environmental Modelling and Software*, Vol. 15, pp. 101–124.

Maniezzo, V. (1994). "Genetic evaluation of the topology and weight distribution of neural network." *IEEE Transaction of Neural Network*, Vol. 5, No. 1, pp. 39–53.

Martin, R. L., and Oeppen, J. E. (1975). "The identification of regional forecasting models using space-time correlation functions." *Transactions of the Institute of British Geographers*, Vol. 66, pp. 95–118.

Montana, D. J., and Davis, L. (1988). "Training Feed-Forward Neural Networks using genetic algorithms." In: Kufmann, M. (Ed.), *11th International joint conference on artificial intelligence*, San Mateo, CA, Vol. 1, pp. 762–767.

Montanari, A., Shoemaker, C. A., and van de Giesen, N. (2009). "Introduction to special section on Uncertainty Assessment in surface and Subsurface Hydrology: An overview of issues and challenges." *Water Resources Research*, Vol. 45, W00B00,doi:10.1029/2009WR008471.

Murphy, A. H., Hsu, W. R., Winkler, R. L., and Wilks, D. S. (1985). "The use of probabilities in subjective quantitative precipitation forecasts: some experimental results." *Monthly Weather Review*, Vol. 113, pp. 2075–2089.

Nasseri, M., Asghari, K., and Abedini, M. J. (2008). "Optimized scenario for rainfall forecasting using genetic algorithm couplet with artificial neural network." *Expert Systems with Applications*, Vol. 35, pp. 1415–1421.

Navone, H. D., and Ceccatto, H. A. (1994). "Predicting Indian monsoon rainfall: A neural network approach." *Climate Dynamics*, Vol. 10, pp. 305–312.

Neal, R. M. (1992). "Bayesian training of backpropagation networks by the hybrid Monte Carlo method." *Technical Report CRG-TR-92-1*, Department of Computer Science, University of Toronto, Toronto.

Nelsen, R. B. (2006). *"An introduction to Copulas."* 2nd Edition, Springer, New York, NY.

Nguyen, V. T. V., McPherson, M. B., and Rousselle, J. (1978). "Tech. Memo. 35." *Urban Water Resources Research Program*, American Society of Civil Engineering, New York.

Obeysekera, J. T. B. Tabios III, G. Q., and Salas, J. D. (1987). "On parameter estimation of temporal rainfall models." *Water Resources Research*, Vol. 23, No. 10, pp. 1837–1850.

Owen, D. B. (1956). "Tables for computing bivariate normal probabilities." *Annals of Mathematical Statistics*, Vol. 27, pp. 1075–1090.

Panofsky, H. (1949). "Objective weather-map analysis." *Journal of Applied Meteorology*, Vol. 6, pp. 386-392.

Pfeifer, P. E., and Deutsch, S. J. (1980). "Identification and interpretation of first order space-time ARMA models." *Technometrics*, Vol. 22, pp. 397–408.

Phanartzis, C. A. (1979). "Rainfall prediction, progress report." *Wastewater Program*, City and County of San Francisco, CA.

Poulin, A., Huard, D., Favre, A. C., and Pugin, S. (2007). "Importance of tail dependence in bivariate frequency analysis." *Journal of Hydrologic Engineering*, Vol. 12, No. 4, pp. 394–403.

Press, W.H., Flannery, B.P., Teukolsky, S.A. and Vetterling, W.T. (1988). *"Numerical Recipes in C. The art of scientific computing"*. Cambrige University Press, UK.

Pun, W. K., Wong, A. C. W., and Pang, P. L. R. (2003). "A review of the relationship between rainfall and landslides in Hong Kong." *Proceedings of 14th Southeast Asian Geotechnical Conference*, Vol. 3, pp. 211–216.

Rabufetti, D., and Barbero, S. (2005). "Operational hydro-meteorological warning and real-time flood forecasting: the Piemonte Region case study." *Hydrology and Earth System Sciences,* Vol. 9, No. 4, pp. 457–466.

Ramirez, J., and Bras, R.L. (1985). "Conditional distributions of Neyman-Scott models for storm arrivals and their use in irrigation control." *Water Resources Research*, Vol. 21, No. 3, pp. 317–330.

Ramirez, M. C. V., Velhob, H. F. C., and Ferreira, N. J. (2005). "Artificial neural network technique for rainfall forecasting applied to the Sao Paulo region." *Journal of Hydrology*, Vol. 301, pp. 146–162.

Renard, B., Kavetski, D., Kuczera G., Thyer, M., and Franks, S. (2010). Understanding predictive uncertainty in hydrologic modeling: The challenge of identifying input and structural errors." *Water Resources Research*, Vol. 46, W05521, doi:10.1029/2009WR008328.

Rodriguez-Iturbe, I. (1986). "Scale of fluctuation of rainfall models." *Water Resources Research*, Vol. 22, No. 9S, pp. 15S–37S.

Rodriguez-Iturbe, I., Gupta, V. K., and Waymire, E. (1984). "Scale considerations in the modeling of temporal rainfall." *Water Resources Research,* Vol. 20, No. 11, pp. 1611–1619.

Rodriguez-Iturbe, I., Cox, D. R., and Isham, V. (1987a). "Some models for rainfall based on stochastic point processes." *Proceedings of the Royal Society London,* Ser. A, Vol. 410, pp. 269–288.

Rodriguez-Iturbe, I., Febres de Power, B., and Valdes, J. B. (1987b). "Rectangular pulses point process models for rainfall: analysis of empirical data." *Journal of Geophysical Research,* Vol. 92, No. D8, pp. 9645–9656.

Salas, J. D., Delleur J. W., Yevjevich V., and Lane W. L. (1980). "Applied modelling of hydrologic time series." *Water Resources Publications,* Littleton, CO.

Salas, J. D., Tabios, G., and Bartolini, P. (1985). "Approaches to Multivariate Modeling of Water Resources Time Series." In *special AWRA monograph, Time Series in Water Resources,* K.W. Hipel (ed.).

Salvadori, G., De Michele, C., Kottegoda, N. T., and Rosso, R. (2007). *"Extremis in Nature. An approach using Copulas."* Water Science and Tehnology Library, Vol. 56, Springer.

Sarle, W. S. (1995). "Stopped training end other remedies for overfitting." *Proceedings of the 27th Symposium on the Interface of Computing Science and Statistics,* pp. 352–360.

Serinaldi, F. (2008). "Analysis of inter-gauge dependence by Kendall's τ_K, upper tail dependence coefficient, and 2-copulas with application to rainfall fields." *Stochastic Environmental Research and Risk Assessment,* Vol. 22, No. 6, pp. 671–688.

Serinaldi, F. (2009). "Copula-based mixed models for bivariate rainfall data: an empirical study in regression perspective." *Stochastic Environmental Research and Risk Assessment,* Vol. 23, pp. 677–693.

Seo, D. J. and Smith, J. A. (1991). "Rainfall estimation using rain gauges and radar: A Bayesian approach." *Journal of Stochastic Hydrology and Hydraulics,* Vol. 5, No. 1, pp. 17-29.

Seo, D. P., Perica, S., Welles, E., and Schaake, J. C. (2000). "Simulation of precipitation fields from probabilistic quantitative precipitation forecast." *Journal of Hydrology,* Vol. 239, No. 1–4, pp. 203–229.

Sklar, A. (1959). "Fonction de repartition à n dimensions et leurs marges". *Publications de Institut de Statistique Université de Paris,* Vol. 8, pp. 229–231.

Sirangelo, B., Versace, P., and De Luca, D. L. (2007). "Rainfall nowcasting by at site stochastic model PRAISE". *Hydrology and Earth System Sciences*, Vol. 11, pp. 1341–1351.

Thielen, J., Bartholmes, J., Ramos, M. H., and de Roo, A. (2009). "The European Flood Alert System – Part I: Concept and development." *Hydrology and Earth System Sciences*, Vol. 13, pp. 125–140.

Thyer, M., Kuczera, G., and Wang, Q. J.(2002). "Quantifying uncertainty in stochastic models using the Box-Cox transformation." *Journal of Hydrology*, Vol. 265, pp. 246–257.

Todini, E. (2001). "Bayesian conditioning of radar to rain-gauges." *Hydrology and Earth System Sciences*, Vol. 5, pp. 225–232.

Tohma, S., and Igata, S. (1994). "Rainfall estimation from GMS imagery data using neural networks." In W. R. Blain and K. L. Katisfrakis (Eds.), *Hydraulic engineering software V,* Southampton, UK: Computational Mechanics. Vol. 1, pp. 121–130.

Toth, E., Brath, A., and Montanari A. (2000). "Comparison of short-term rainfall prediction models for real-time flood forecasting." *Journal of Hydrology*, Vol. 239, pp. 132–147.

Versace, P., and Capparelli, G. (2008). "Empirical hydrological models for early warning of landslides induced by rainfall." *Proceedings of the First World Landslide Forum* (18–21 November 2008, Tokyo), pp. 627–630.

Versace, P., Sirangelo, B., and De Luca, D. L. (2009). "A space-time generator for rainfall nowcasting: the PRAISEST model." *Hydrology and Earth System Sciences*, Vol. 13, pp. 441–452.

Villarini, G., Serinaldi, F., and Krajewski, W. F. (2008). "Modeling radar-rainfall estimation uncertainties using parametric and non-parametric approaches." *Advances in Water Resources*, Vol. 31, No. 12, pp. 1674–1686.

Yano, K., and Senoo, K. (1985). "How to set standard rainfalls or debris flow warning and evacuation." *Sabo Symposium* (SEDD Japan), pp. 451–455.

Yoo, C., Jung K. S., and Kim T. W. (2005). "Rainfall frequency analysis using mixed gamma distributions: evaluation of the global warming effect on daily rainfall." *Hydrological Processes*, Vol. 19, No. 19, pp. 3851–3861.

Yu, Y. F., Lam, J. S., Siu, C. K., and Pun, W. K. (2004). "Recent advance in Landslip Warning System." *Proceedings of the first day Seminar on Recent Advances in Geotechnical Engineering* (Hong Kong Institution of Engineers Geotechnical Division), pp. 139–147.

Yu, Z.W. (2001). "Surface interpolation from irregularly distributed points using surface spline, with Fortran program." *Computer and Geosciences*, Vol. 27, pp. 877–882.

Zawadzki, I. (1987). "Fractal structure and exponential decorrelation in rain." *Journal of Geophysical Research*, Vol. 92, No. D8, pp. 9586–9590.

INDEX

human, 57
humidity, 66, 68
hybrid, 157
hypothesis, xv, xvii, xxv, 4, 14, 16, 22, 23, 24, 40, 57, 80, 91, 125, 129, 131, 132, 133, 138, 139, 147, 148, 149
hypothesis test, xvii, 132, 138

I

ideal, 74
identification, 3, 14, 15, 157
identity, 59
illusion, 5
imagery, 160
images, 41, 57
independence, 109
India, 57
individuation, 22, 35
infrastructure, 2
initial state, 65, 73, 74, 75
intervention, 2
inversion, 11, 133
irrigation, 158
Israel, 154
issues, 157
Italy, 1, 28, 79, 153

J

Japan, 160

L

laws, 5
lead, 27, 28
leadership, 72
learning, 60
learning process, 60
linear function, 24, 27, 29, 35, 40, 58, 59, 84
linear model, 152

M

magnitude, 3
marginal distribution, xxiii, 15, 20, 43, 48, 50, 51, 52, 54
mass, 67
matrix, xix, xxvi, 23, 25, 109, 143, 144, 145, 146
measurement, 2, 80
measurements, 62, 72, 73
median, xxiv, 107, 116
memory, 22, 23, 27, 29, 35, 39, 40, 41, 90
methodology, ix, 4, 5, 11, 22, 35, 48, 58, 62, 63, 91, 94
modelling, 1, 13, 27, 66, 71, 91, 154, 157, 159
models, ix, xiv, xxiv, 3, 4, 6, 9, 11, 13, 14, 15, 16, 18, 22, 30, 44, 47, 50, 57, 60, 63, 65, 66, 67, 70, 71, 73, 75, 77, 79, 82, 93, 94, 151, 152, 153, 155, 156, 158, 159, 160
modules, 76, 77, 94
moisture, 72
Montana, 62, 157
Monte Carlo method, 63, 157
Morocco, 153
multidimensional, 102
multivariate analysis, 30, 48, 51, 52, 109, 125
multivariate distribution, 18, 50, 82

N

National Research Council, 72
NATO, 152
nervous system, 57
network density, 31, 37, 40
neural network, 153, 154, 156, 157, 158, 160
neural networks, 154, 156, 160
neurons, 57
New England, 151